Henry J.B.M. Duke of Rutland

The Trout

Henry J.B.M. Duke of Rutland

The Trout

ISBN/EAN: 9783744793919

Printed in Europe, USA, Canada, Australia, Japan

Cover: Foto ©berggeist007 / pixelio.de

More available books at **www.hansebooks.com**

THE TROUT

BY THE

MARQUESS OF GRANBY

WITH CHAPTERS ON

BREEDING
BY COLONEL F. H. CUSTANCE

COOKERY
BY ALEXANDER INNES SHAND

ILLUSTRATED BY J. E. GRACE AND FROM PHOTOGRAPHS

NEW EDITION

LONGMANS, GREEN, AND CO.

39 PATERNOSTER ROW, LONDON
NEW YORK AND BOMBAY

1899

PREFACE

THE design of the *Fur, Feather, and Fin Series* is to present monographs, as complete as they can possibly be made, on the various English birds, beasts, and fishes which are generally included under the head of Game.

Books on Natural History cover such a vast number of subjects that their writers necessarily find it impossible to deal with each in a really comprehensive manner ; and it is not within the scope of such works exhaustively to discuss the animals described, in the light of objects of sport. Books on sport, again, seldom treat at length of the Natural History of the creatures which are shot or otherwise taken ; and, so far as the Editor is aware, in no book hitherto published on Natural History or Sport has information been given as to the best methods of turning the contents of the bag to account.

Each volume of the present Series will, therefore, be devoted to a bird, beast, or fish. Their origin will be traced, their birth and breeding described, every known method of circumventing and killing them—not omitting the methods employed by the poacher—will be explained with special regard to modern developments, and they will only be left when on the table in the most appetising forms which the delicate science of cookery has discovered.

ALFRED E. T. WATSON.

CONTENTS

TROUT FISHING

By the Marquess of Granby

TROUT BREEDING

By Colonel F. H. Custance

THE COOKERY OF THE TROUT

By Alexander Innes Shand

ILLUSTRATIONS

By J. E. Grace, and from Photographs by
Herbert Bickerton, William Penn, and others

(*Reproduced by the Swan Electric Engraving Company*)

Various Diagrams in the Text : those in the Chapters
on Breeding by C. A. Calvert

TROUT FISHING

BY THE

MARQUESS OF GRANBY

CHAPTER I

FLY FISHING FOR TROUT

THE year of Jubilee produced a wonderful crop of
statistics showing the marvellous increase in the
wealth, power, and influence of England during Her
Gracious Majesty's reign. Every loyal subject rejoices
that this is so, though the unfortunate agriculturist
may not be tempted to vouch for the accuracy of the
statement as regards the particular industry in which
he is engaged ; but in no trade, in no industry, has a
more remarkable development taken place than in
that connected with the supply of articles necessary
for the various sports and athletic competitions accounts of which now occupy so much space in the
daily papers, and add so much interest to the lives of
vast numbers of our fellow-countrymen. Take, for
example, the scene at the Oval on an August Bank
Holiday, when Surrey and Notts are engaged in a
County cricket battle, or at Lord's when the matches

B 2

Australia *versus* England are in progress, or indeed,
look anywhere where good cricket is being played.
Then as to football, what an extraordinary concourse
of people flock to see the final tie of the Association
Football Cup! I saw some 65,000 gathered together
at the Crystal Palace for this purpose in the April of
1897. Notice, moreover, that every Saturday during
the winter, when the Association Football teams are
competing, there are generally many more than
100,000 people looking on at that competition alone,
to say nothing of the crowds attending the ordinary
matches. Or turn to Golf; it is but a very few years
ago that the game was played in England only at
Blackheath, Wimbledon, Westward Ho! and one or
two other places; but now all over the country golf
links are to be found. No watering-place, no health
resort, no town of any importance or progressive
aspirations is without this adjunct to the happiness
of its inhabitants and visitors. In fact, it is not too
much to say that a town labours under a considerable disadvantage, if it be dependent on the advent
of visitors, should no golf links be available in its
immediate vicinity.

So with all other kinds of athletics, public interest in them has grown enormously during the past
decade; in fact of very recent years, sport in every

form has developed in Great Britain in a most striking degree; but I venture to think that no sport or pastime has attracted a greater number of votaries than has fishing, whether it be for coarse fish or trout. Let him who does not credit this go, on almost any day between May and August, to one of the main London stations when an early train is leaving for the districts through which Kennet, Test, Lea, Mimram, Chess, or other trout-yielding river gently flows. His scepticism touching the attractiveness of the 'gentle art' as practised in this year of grace must then surely be abandoned. If it be moderately favourable weather, there will almost to a certainty be there discovered not a few men clad in rough workmanlike clothes, bearing fishing bags and rods, who are anxiously wondering whether for once the fates are going to be propitious and send them that most essential—and also for some reason or other rarest of all the elementary conditions of fishing, an up-stream breeze. Where ten years ago there might be found one such angler, now twenty have appeared.

Or let our doubting friend go to the Midlands of England and seek the banks of Wye, Dove, or Derwent, and there, on every possible occasion, such as a Bank Holiday or a Saturday during the fishing season, he will discover people who have come long distances to

try to lure from those lovely streams the much-fished-
for and wonderfully wary trout who dwell therein.
On enquiry he will discover that the number of
tickets now sought for the open waters is largely in
excess of that for which demand was made but a few
years since. So it is in the North ; and, in fact, turn
to whatever part of England you will, if trout be
found there, and reasonable facilities for preserving
the rivers be afforded, you may be sure that endless applications to buy or rent those streams will be
made.

In truth, the owner of a good trout river, should
he be no fisherman himself—and then I should
sincerely pity him—has now a safe and certain
source of income ; indeed, should he be the fortunate possessor of a part of one of the favourite
southern streams, he can ask and obtain a price for his
fishing which but a few years since it would have
been considered an act of lunacy to mention. One
may be fairly asked why this sudden demand for
trout fishing should have taken place, and why the
followers of old Isaac's craft should have all at once
become so numerous ? In the North there have
always been many more anglers than in the South ;
more literature on the subject of trout fishing has appeared there than elsewhere, and therefore it is with

the more southern portions of England that we must now deal in pursuing these considerations.

Charles Kingsley undoubtedly contributed largely towards the popularising of trout fishing. It would be scarcely possible to read his delightful descriptions of Devonshire and Hampshire without feeling a desire to attempt, in however small a degree, some such piscatorial deeds as those recounted in 'Chalk Streams Studies' and other works of that great exponent of healthy, honest, English life and sport. Moreover, the facilities now afforded by the railways allow a busy man to get away from his town work for the inside of a day, and in two or three hours or less he can find himself by Itchen or Kennet side, should he be fortunate enough to have fishing thereon. The train service will enable him to return to London the same evening, so that only one day will have been taken from his working time. But a few years ago this would not have been practicable, owing to the then inferior railway arrangements. This of course applies equally to any of the great centres of population in the country as to London.

Let us therefore admit, as an accepted fact, that trout fishing has become vastly more popular than it was twenty years ago, and that the number of its adherents has largely increased.

The next point to be considered is, what particular method of fishing is now usually practised. Is it fly fishing? If so, do people fish up or down stream? Do they use wet or dry fly? Or do they sometimes attempt to attract their victims by means of the well-scoured worm?

It may, I think, be safely averred that nearly every one of the great army of anglers nowadays fishes *up*-stream with the dry fly; that is, whenever such method be practicable; and also, that on hardly any river, whether it be open to the public or strictly preserved private water, is the use of anything else than artificial fly allowed.[1]

Let us now enquire what it is that constitutes the great charm of fly fishing, and especially the use of the dry fly; and further let us penetrate somewhat into the mysteries of this art—for it surely deserves such a designation.

With regard to the reason of the attractiveness of this form of fishing, it must be admitted that it is infinitely more satisfactory, from a sporting point of view, to mark a good trout lying near the top of the water in some almost painfully clear and stilly flowing stream, and determine to secure him, if by any fair

[1] I am here alluding mainly to the trout rivers of England. In Scotland, and also in parts of England where rough quick streams exist, wet-fly down-stream fishing is more generally prevalent.

agency such end may be obtained, than to seek out
some rough piece of water, and there, by throwing
blindly, and without any idea of the size or nature
of fish that may take your drowned fly, cast on the
'chuck and chance it ' principle.

In the first case you pit your skill and piscatorial
and entomological knowledge against the trout's
natural instincts and sharpness of eye—how keen
this vision is only fishermen know; whereas when
blindly casting into runs and rough water you have
little or no idea of what is happening to your fly,
or what the fish thinks of it and you ; unless by
chance some unwise troutlet hooks itself, or what
is more probable only gets pricked, and hurriedly
retires into the security of the nearest weed bed.

To prevent any misconception, however, I would
point out that dry-fly fishing can only be followed
under certain conditions, and that excellent sport
and capital fun are to be obtained by the other forms
of fly fishing, as will be explained elsewhere in this
volume.

The methods of dry-fly fishing can be divided
into three heads : —

First, fishing for trout 'standing,' *i.e.* fish which
are near the top of the water looking about for flies,
or other floating food.

Secondly, fishing for fish which are rising, but

which, owing to a slight ripple on the water, or some other cause, cannot be accurately seen.

Thirdly, fishing over likely places where trout are not observed to be either ' standing ' or rising.

It would seem that numbers one and two are so closely allied that they must be described together ; for though there are certain differences in the manner in which trout should be approached under such circumstances, so much similarity exists that separate treatment of the methods would be superfluous.

Let us then, through the instrumentality of that most marvellously vivid medium—Imagination— transport ourselves to the bank of some crystally clear stream, place it where you will, in Hampshire or Buckinghamshire, in Hertfordshire or Derbyshire, and there let us in fancy try whether we can wile from the depths certain big trout which we know dwell therein.

In what month shall we suppose ourselves to be ? Let it be May, and towards the third week. Then shall we get the benefit of many things to help us to enjoy our sport, both as regards the pleasure of the eye, and the success of our undertaking. For he is no true angler who does not to the depths of his soul appreciate the beauty of the surroundings which must at all times be with him when by the banks of any river, though it may flow through what to the casual observer would seem to be but a dull and uninteresting country. As soon as ever he is actually on the water, so to speak, he must be conscious, if worth his salt as a fisherman, of all that Nature has placed before him for his notice. Though the country down to the river's brim be flat and uninviting, still, as the water's influence is felt, there appear beautiful creations for his attention, and if he be wise, for his careful study.

The masses of orange celandine; the iris, standing in bold relief, with its yellow spikes of flowers uplifting themselves among their graceful leaves; the forms of the myriad wild flowers upon the banks, should gladden both eye and mind; while, should any coppice or wood approach the river, then at various seasons the starwort, herb robin, bluebell, forget-me-not, anemone, wild mignonette, and, in the North, the wild geranium, lend their gracious aid to the beauty of the scene, by the carpet of deftly blended colours which they spread over the land.

And what shall be said of the bird-life which is always to be found by the riverside? Nightingale, peewit, redstart, summer snipe, kingfisher, blackcap, and a hundred other kinds, all seem to select a stream as their favourite haunt: and the value added to the charm of a day's trout fishing by the presence of these sweet songsters, ought to be of inestimable worth to him who can in any degree appreciate the beauty of Nature's works.

But however devoted a naturalist our fisherman may be, he cannot help feeling anxious to begin operations, especially as this is his first day's fishing of the season. So let us set him to work so soon as may be ; as not only does the day seem favourable, but there is much promise in what is happening only a few yards above the spot where we are standing : for sundry circles have appeared just below a small weed bed, showing that *some* fish, at any rate, are beginning to ' move.' Moreover, there is — oh fortunate day !—a gentle wind blowing up-stream ; therefore no forcing of the fly will be necessary, and no compulsory cross or down stream fishing need be undertaken. Indeed, our angler seems to have 'happened' on an ideal day for his purpose, though, as fishermen know too well, ideal days as regards weather very often turn out singularly unsatisfactory ones with respect to sport. Still, pleasant atmospheric conditions count for much when fishing, though the bag may not be so heavy as anticipated when the rod was first put together in the morning.

But while these reflections have been proceeding,

the ten-feet six-inches-long rod has been made ready,
the reel fixed, and a delicate cast attached to the line.
And now the all-important question has been reached,
what fly shall first be tried on this eventful day?
Shall it be olive-dun, olive-quill-gnat, red-quill,
alder, or what? Much must of course depend on
the part of England in which our imaginary stream
is placed; but as it is necessary to generalise, let
us suggest to our fisherman that an olive-dun tied
with a slight gold twist on the body is as likely a
fly as any to attract a trout. So on with it at once.
Delicately fashioned, with wings gaily cocked, it is
as like as possible to the small duns which are seen
dancing about in the sunshine, though it does not
appear that as yet there are many floating on the
water.

Before launching it on its voyage of deception,
however, one more ceremony has to be observed; not
in this instance a libation, but a very slight touch
of refined scentless paraffin oil given to the body and
hackles of the fly, so as to make it float better, and in
consequence reduce to a minimum the necessity of
constant casts in the air to dry it.

And now having completed all preparations, and
having made certain that the cast was thoroughly
soaked before being attached to the line, let us

accompany our angler to the spot where he proposes
to start operations.

It must be here premised that in this particular
river the trout are carefully preserved, are large in
size, and wonderfully well educated as regards taking
care of themselves. Therefore much caution must be
observed when proceeding against them. But see !
our friend, after standing still for some moments,
during which time he has been steadily gazing up-
stream, has sunk on to his knees, and, keeping his rod
low and behind him, crawled a few yards towards the
bend of the river where the first rises were noticed.
At least two fish are moving there now, the nearer
one being apparently the larger; which is fortunate,
as thus no disturbance of the water is probable,
which would have been the case had the reverse
happened and our fisherman been obliged to try for
the smaller trout first. If hooked, his struggles
would in all probability have scared the other, bigger,
fish away ; whereas it does not matter if the little one
be alarmed and retires.

Before the first cast is made it may be worth
while to mark carefully the peculiarities of the water
wherein are the rising trout. They are lying in a
kind of small angle of the river close to the side farthest
from the angler. A couple of yards above them is a

A DIFFICULT STALK

half dam, a plank which slightly turns the stream
towards the bank whereon the fisherman crouches, and
they are rising close to the edge of this small run.
Overhead are some pollard willows, which, however,
do not much interfere with the casting. The bank is
slightly raised on the side from which our friend must
try for the fish, and he can, by kneeling behind this
rise in the ground, secure an admirable and much
appreciated shelter from the marvellously acute sight
of the trout. But look ! there he is again, so try for
him at once, while he is in the mood. There ! your
first cast was a trifle below him, and as you did not allow
sufficiently for the swirl of the stream between the
trout and yourself, the fly dragged somewhat, and did
not float properly at all. That's better and sure
enough up he comes. You have him ; and now mind
what you are about ! As I thought—he is off, to
bury himself amongst the weeds at the corner of the
dam where it touches the bank under the shelter of
those flags. Put whatever strain you dare on him,
as, if he reaches the haven where he would be, you
will probably lament the loss of both fish and fly ;
the weeds are tenacious hereabouts. But it is needless to attempt to school our angler, who knows what
he is doing, and will surely account for this fish if
any one can. And so it comes to pass that, in a few

moments, the weed dangers having been overcome,
the landing net is carefully slid under the trout, and a
beauty of a pound and a half—or whatever weight
the imagination chooses to create—is triumphantly
placed on the bank.   The first fish of the season ! a
very thrilling moment for any really enthusiastic
fisherman !

Let us accompany the angler for a little while
longer, as he warily and very slowly moves up-stream
in quest of another quarry ; and pray observe that, just
round the corner of the next little bend, the surface of
the water is gently touched by the soft breeze which,
though of a perfect strength and direction for fishing,
sufficiently ruffles the stream to prevent one from
seeing the trout in it, or observing the exact place
where they rise.   So, this particular bit of the river
being generally accepted as one never to be heedlessly
neglected, it must be fished on the present occasion
in the third manner indicated at the beginning of this
chapter : that is to say, the fisherman, not being able
to see either trout or rise, must bring into action his
knowledge of the gentle art, as well as, should he
possess it, his previous experience of the water, and
put his fly over those portions of the stream which he
thinks are most likely to hold a trout.

And you will please to remark that the fly is

not cast wildly at hazard on to the water, into mid-
stream, or anywhere else. It is a useless performance, that of fishing in what Canon Kingsley calls
'skip-jack broad.' It is thrown deftly to the bank's
side ; it floats down near the reed edges ; it sails
jauntily over the tiny ripples which lap the weed
beds fringing the river's banks. Fish that mean to
take the fly are fish which are feeding, and they know
well that from the bushes, flowers, grasses, reeds and
trees growing by the river's sides comes most of the
surface food that is supplied to them.

The water insects which, when hatched out, crawl
up the stems of the grasses and rushes to expand and
perfect themselves in the sun's warmth ; the various
beetles, ground bees, caterpillars, and suchlike, come
from the thick growths upon the banks and fall therefrom into the water, obviously not into the middle of
the stream, but close to the edge. Every overhanging
branch, however small, of willow, alder, thorn, or other
tree and bush, provides some tasty morsel which drops
into the river near the banks. Every stem and stalk
contributes its portion of attractive food for the fishes'
delectation ; and therefore it is only very young and
foolish troutlets who course about in the centre of
the stream when looking for their meals, and no
one heeds *them*. The older, larger, and more worthy

c

trout lie quietly in their selected haunts, and take the goods the gods provide for them as they drop lightly from grass and reed, stalk, bush, or tree ; and it is these more experienced fish of which our angler is in quest.

And, indeed, he seems to be in luck's way to-day, for between that weed bed which is just appearing in mid-stream and the nearer bank, his assiduity is apparently rewarded. Surely that is a good fish which has just been hooked. Ye gods ! that was an up-stream rush indeed ! But the aforementioned weed-growth looks dangerous, and by every power that torments the angler's soul, the fish is into it ! Well, can he be got out ? Possibly, but not certainly ; for he has taken a quick turn downwards after making good his endeavour, and such manœuvres are often fatal—not to the fish. Anyhow, the trout will not come out of his own accord, and the fly is fast in something, whether it be fish or weed ; so the fisherman must wade in, and with the handle of the net see whether the trout, if he be still on the fly, can be induced to move, instead of burrowing about like an unseemly rabbit. So, is he still there ? What—gone ? Well, it can't be helped. If the fly cannot be disentangled it must be broken off, and a fresh one and a new length of gut attached. These things will happen, and if fishing were reduced to a certainty where would be its charm?

But we will not follow our angler further. We will wish him a sufficient basket and a full day's sport, and turn for a short time to the consideration of the methods employed for the inveiglement of these fish.

In the first place, it will be noticed how carefully all the apparatus was handled, the rod put together, the casting line tested, and the fly chosen and attached. Any one who is an adept can tell almost at a glance, when watching a fisherman put up his rod, whether he is an old hand or the reverse. The beginner, should he happen, on first reaching the riverside, to notice that a fish or two is rising, will in all probability begin hurriedly to get his rod ready, looking all the time up and down stream as if he feared the trout would run away before his preparations were completed. The old hand will very likely turn his back on the water and not trouble himself about anything except the business immediately before him, viz. that of getting his tackle ready.

But the reader may ask, what does it matter whether these things are done deliberately and slowly, or more hurriedly and rapidly? It matters this much: that in the one case the joints of the rod are properly adjusted, so that no unfastening or loosening is likely ; the casting line is sufficiently soaked, examined, and straightened, and the fly carefully chosen both as to size and character ; whereas, in the other case—that of the novice—the rod, if a split-cane one, is very likely strained in the adjusting, or the fastenings are not properly secured, thus necessitating further attention later in the day, generally at the most inconvenient moment. The reel line has probably not been over-hauled prior to leaving home, and having been put away when damp at the end of the previous year, it is rotten, and breaks with the first heavy fish ; the casting line is put on in a brittle, 'kinked,' condition, the hurry of the occasion causing our tiro to neglect, or to observe in a very partial degree, that most essential preliminary of thoroughly soaking it before attempting to use it ; while, as regards the fly, one may perhaps be selected which is not that most similar to the natural insect on the water, of which a copy is in the angler's fly-book. Moreover, in his excitement, the young fisherman will very likely get his cast into a muddle, and thereby defeat the very object he had in view—that of making everything ready at great speed.

It will be readily seen, therefore, that hurry is most undesirable when beginning a day's fishing.

In the second place, it will have been remarked that our expert did not begin to cast the very moment he had completed his preparations, but stood for a while looking both up and down stream to see where any fish were moving, so as accurately to 'locate' their whereabouts, and obviate the necessity of beginning to fish without knowing as exactly as possible where he had better place his fly. He concluded on the whole that it would be better to start with the trout which he saw rising in the river above him than to go down-stream some way, making a wide détour into the fields so as to avoid disturbing the water, and he therefore decided to 'go for' the fish whose capture has been described. Having thus made up his mind, he walked very carefully up-stream some way off the water, and having arrived within a measurable distance of his object, he approached the river, and, sinking on his knees, judged carefully the space which intervened between himself and the rising trout. How he arrived at the right spot for so doing may be explained by saying that he had carefully marked the place by a particular willow-tree just above the doomed fish. It need perhaps hardly be added that our angler was equipped with a pair of light waders—for rheumatism is a thing to be avoided as much as possible.

Having arrived thus far in his stalk, the fisherman began to cast, not wildly, but circumspectly, being careful not to *over*cast at first. As has been related, his first effort was short—a sound fault, if fault it be —and having thus, like a good bowler, 'found out his length,' the second attempt was altogether successful.

The next move was ·to prevent the trout from reaching his hover, or home, in the weeds near the dam, and this being done, he gradually worked him down-stream and landed him several yards below where he was hooked. Again, the landing operation was effected scientifically. The net was not taken off its string or strap till the fish was tired out and practically killed, and no wild dives were made with it. At the right moment it was placed gently in the water *below* where the trout was, and the capture easily effected. The fish was then at once killed, either by knocking its head against the toe of his wader, or by breaking its neck by inserting the thumb in the upper jaw and bending the head back till the neck-bone broke. Thus the whole affair was conducted skilfully, deftly, and in order. No undue excitement, no confusion, prevailed.

What would in all likelihood have occurred if, instead of an experienced fisherman, it had been a young and unpractised hand who had essayed to kill

NEARING THE END

this fish? As has been suggested, he would have probably begun with an unsafe reel line and half-soaked gut, either or both of which might break at any moment. He would very likely have forgotten to approach the fish with the necessary caution, and, by showing himself or his rod unduly, have frightened him away before ever a cast was made. If by chance our tiro *did* succeed in hooking the fish, then the improperly prepared tackle would perhaps have given way; or if by some lucky chance it held, he would not have calculated on the first fierce rush of the trout, and would have allowed him to reach his weed-surrounded home, when it is long odds against any more having been seen of either fish or fly. Then, instead of rejoicing over the first success of the year, lamentation and woe would have been that angler's portion.

Probably it will be said, with perfect justice, how in the world can anyone become a 'complete angler' all at once? It is absurd to suggest it, and no one is better aware of this than the writer. But it is by showing, as best one may, how matters piscatorial should be undertaken, that those who are desirous of becoming skilled fishermen may attain their object; and for this reason the points of difference between the methods of procedure, as practised by one who has, so to speak, taken his fishing degree, and by one who has only just begun his angler's career, have been described.

But let us suppose that, instead of a gentle *up*-stream breeze, a strongly blowing *down*-stream one has greeted our friend when he reached his river. What is he to do? There is too much strength in the wind for him to attempt to fish up-stream, except in certain tiny bends of the river where such a proceeding is possible. It is open to him, if he wants to fish a fair amount of the water, to try to work it either by throwing directly across and slightly down, or to boldly make up his mind to fish down-stream altogether, on the ' chuck and chance it ' principle aforementioned.

There is a nice ripple on the surface of the stream, should he decide on the latter course, and so he does. With the stronger wind now blowing, a rather larger fly than that which would be selected for a calm day should be used. What shall it be? A large red-quill-gnat, an alder (although rather early in the season for it), or a March-brown? On the whole, perhaps an alder is likeliest, and therefore let that be the fly chosen.

Keeping as much out of sight as possible—for this precaution must never be neglected when fly fish-

ing for trout, even though the state of the water and
light might seem to render it unnecessary—our
angler begins to fish carefully, with a longish line, the
stream on his own side of the river.   Gradually, little
by little, he works across the water till he is fishing
close to the farther bank, when a faintly perceptible
break in the wind-touched river and a slight tug pro-
claim that a trout has 'come at' his fly just under
water and has missed hooking itself.   And why was
this?   The strike was made quickly and fairly, but
the fish was only just pricked.   The reason is a
simple one.   When fishing down-stream many more
fish are touched and not hooked than when fishing
up-stream ; for, in striking, the hook is obviously
pulled away from the fish's mouth, as all fish lie with
their heads up stream instead of down.   That is one
difference in the two methods.   Another is, that
whereas in up-stream fishing with a dry fly the great
object is to let the fly float on the water as quietly
and naturally as possible, in down-stream fishing a
slight movement should be imparted to it, presumably, as the fly is either partially or altogether under
water, in order to imitate the struggles of a drowning
insect.   When fishing up-stream, one can generally
manage, unless the fish is a very large and obstinate
one, to prevent it from going far above where it was

hooked, and so spoiling the water there. On the other hand, when fishing down-stream, it is almost impossible to pull a fair-sized trout up against the current. All that can be done is to get down level with it as quickly as may be, and endeavour to check, with all the strain one dares to put on the tackle, any down-stream rushes which the victim may attempt to make.

While we have been considering these various matters our fisherman has been steadily working his way down-stream, trying, with untiring perseverance, every likely nook, and the edge of every weed-bed which he fancies may be the haunt of a lusty trout.

Nor are his efforts uncrowned with success ; for see ! he has even now risen and hooked a fish which came at him boldly and fastened well. Notice how— not allowing the trout an inch more line than is imperative—he hurries down, reeling up for dear life while he goes, so as to get abreast of the fish and prevent his going further down the river, thus disturbing water as yet untried. There is not much doubt as to the result, so soon as the angler has got level with the fish. There are not many weeds in mid-stream in this part of the river, and the trout is well hooked. Thus the end comes soon, and owing to the skill and

quickness of the angler but little of the water below where the fish was taken has been spoiled.

Fishing is a sport best pursued alone ; though at lunch, or after the day is over, no one is less inclined for solitude, or more desirous of the company of his fellow-creatures than is a fisherman. This, however, is neither lunch nor dinner time, so we will betake ourselves and our observations elsewhere, leaving behind us every piscatorial good wish.

Such are the methods usually adopted when fly fishing for trout in English rivers. To recapitulate them, they are, broadly speaking : (1) fishing up-stream with the dry fly ; (2) fishing down-stream with a wholly or partially sunk fly.

Naturally there are many exceptions to these general rules. Take, for example, a Derbyshire or North-country river in early spring-time, when there is plenty of water, and trout are rising boldly and hungrily. On a favourable day it is well worth the angler's while to put on a couple of flies instead of one only, and to fish either up or down stream as seems best to him. It would be folly on such an occasion to be tied by any hard-and-fast rule condemning one to use only a single fly or to fish solely up-stream.

There are two conditions of mind or body which often affect trout, to which I have, as yet, made no

allusion. The first is when trout are what is called
' tailing.' The second, when they are ' bulging.'

With regard to 'tailing,' it is a performance frequently witnessed in Hertfordshire and Buckinghamshire waters, and in South-country streams generally.
As may be surmised, the phrase 'tailing trout' means
that the fish are plunging their heads into the weeds
or poking about on the bottom of the river, seeking
their food amongst the fresh-water shrimps, grubs,
and similar appetising morsels ; and in comparatively
shallow streams this downward operation causes
the fishes' tails to appear above the surface of the
water. Under circumstances such as these it is no
use whatever trying for them with the dry fly. I
have, nevertheless, frequently seen considerable
execution done amongst trout when in this humour
by those who have had experience of this particular
phase of fish-life. The most general plan is to cast
down-stream with a long line, and to work the fly a
little. As ' tailing' usually takes place in the months
when the alder is about—I fully appreciate Canon
Kingsley's well-known eulogy on that most admirable
insect—I cannot name any fly that is more likely
to prove fatal under such conditions. In fact, I have
often worried a ' tailing' trout into ' catching hold,'
even when fishing up-stream ; nearly always when

using the alder. But far more often the down-stream method is efficacious. I was sent last year a pattern of a small partridge hackle fly which had killed several good trout on the Mimram, all tailing fish. On enquiry I found that the fisherman, a clergyman who, like most of the cloth who are anglers, is a capital performer with the rod, had caught these fish by casting up-stream and letting the fly sink just above where they were feeding. It cannot, of course, be expected that as good a basket can be obtained when trout are 'tailing,' as when they are rising. Nevertheless no one need despair of killing some 'warrantable' fish when he sees not noses, but tails, breaking the surface of the river he is about to exploit.

As I have before mentioned, hackle flies are most useful. In fact, winged ones (always except-ing the alder) are really not worth thinking about in connection with 'tailing' trout. Palmers, Zulus, spiders, all do execution. But the fisherman must not expect that a 'tailer' will be incited to hook himself at once. Endless perseverance will in all probability have to be exercised. For remember that a trout feeding in this fashion is rather like a man enjoying a particularly good dinner, to whom a strange and incongruous dish is suddenly offered

by some-one totally unconnected with the feast.   He
would probably hesitate considerably before accept-
ing the proffered novelty, for fear of either spoiling
the proved excellence of his meal, or from the dread
of some form of practical joke.   And that is just what
trout do when 'tailing.'   They think over the matter
deeply, before venturing to attend to the allurements
of the daintiest hackle fly.   So the fisherman, when
he tries for a 'tailing' fish, must make up his mind
to work hard for him, and then, with luck, he may
eventually be well rewarded for all his time and
trouble.

The second condition I have mentioned is, when
trout are found to be 'bulging.'   Often on reaching
the riverside the fisherman will notice what at the
first glance might seem to be innumerable rises.   On
closer inspection these will prove to be fish taking
what I believe are called the *nymphæ* of the duns or
mayflies just before they reach the surface, there to
hatch out.   When trout are thus employed, they are
very difficult to catch by means of the artificial fly,
however skilfully used.   A stray fish may be induced
to look at a floating dun, but it is a rare occurrence,
and on the whole I don't know any more aggravating
mood to find fish in than that of 'bulging.'   What
makes it worse is that the water is being constantly

moved by the trout hurrying about just underneath the surface to secure the tasty morsels which are rising to the top ; and there must be always a general impression that, with all these fish apparently ' looking about,' by some means or other a few should be secured. But believe me, brother anglers, to succeed when fish are ' bulging ' is an extremely difficult task. Some weird, odd, fancy fly is, I think, then more likely to entice a trout than the regulation dun or alder.

I have as yet said nothing about those most aggravating of all created insects, ' the smuts,' or ' fisherman's curses.' Now, whether the latter nomenclature was applied to them because of the nuisance they are to the angler, or because they cause the wearied fisherman to use strong epithets, I am unable to decide ; all I can say of myself is that many and many a time have they utterly destroyed the remnants of my own temper, and I know of nothing which can compare with them in this respect except the game of golf ! These infinitesimally tiny creatures, when they do appear in force—not that they are even then clearly visible to the eye of man—seem to deprive the trout of the power of attending to the charms of any other insect. After a toilsome hot day's fishing, when but little success has been achieved, how often have I seen trout begin to rise quietly but regularly, and have imagined in my ignorance that the short time still left me would prove not unprofitable ! Then have I fished as carefully and steadily as I could, but not a fish would look at olive-dun, red-quill-gnat, blue-dun, or alder !

On closer examination of the water and its stubborn inhabitants, I have observed the trout rising persistently at some object so absolutely indistinguishable that it was an impossible task to attempt to essay the use of any artificial copy thereof. The fish were 'smutting.' That was the simple explanation, and one which is, as a rule, a final extinguisher to any hopes that might still linger of being able to make up a decent basket before going home. Myriads of insects, no larger than a pin's head—if, indeed, they were as big—were wheeling in the air—millions upon millions of them—and multitudes were dancing on and over the water. Upon these the trout were feeding, and no other fly was then of any avail whatsoever.

It is true that on rare occasions I *have* known an incautious fish to take a small black gnat or a wee dark dun ; but these are exceptions that prove the rule. No, the 'fisherman's curse' when it comes up

in strength, and the fish have made up their minds to
take it, is almost certain to completely and absolutely
defeat the unfortunate angler. And with this somewhat discouraging statement I will take leave of
' smuts' and their evil influence over the trout.

Before quitting altogether the subject of fishing
with a dry fly—or, indeed, with the artificial fly
at all--a few words must be devoted to that most
fascinating and weird insect, the Mayfly.

From the moment the first pioneers of the tribe
are seen, to the day when a few scattered stragglers
are the sole survivors of the vast multitudes who
have fluttered their little hour away and have died
either a natural death, or have been devoured by bird
or fish, the Mayfly is the thought uppermost in
the fisherman's mind, and he will travel far to procure
a day's fishing in waters whereon that fly is found.
Apart from the wondrous tale of its inception, and
curiously ephemeral existence, the Mayfly is always
affording the fisherman some new material for aston-
ishment. One year it will not appear at all, or only
in such insignificant numbers as hardly to merit
recognition, under which circumstances the fish take
no notice of it whatever, and the angler need not
trouble to ' put on' its likeness ; whereas on the same
stretch of water during the previous season the fly

D

may have been 'up' thickly, and many a heavy
basket have been the result.

The Mayfly will perhaps perform this disappearing
trick for a couple of years in succession, and then
for some unexplainable cause it will again dance gracefully on to the scene, and the soul of the fisherman
will once more rejoice.

Again, why should the Mayfly sometimes draw a
hard-and-fast line at certain lengths of a river whereon
they abound everywhere else except upon those particular stretches?   I do not know, and I doubt whether
any one does.   The appearance of the water in both
the favoured and despised portions of the stream may
be identical; the surrounding vegetation of every
description may be absolutely similar and as abundant; and yet in the one place is the Mayfly, and in
the other it is not.   What an interesting insect it is,
and what a strange brief life it has !

Consider for a moment what an amount of trouble
Dame Nature has taken to produce this delicately
fantastic insect, and then think for how long, after it
has reached its perfect form, it is allowed to exist !
Not less than two years elapse from the time when
the eggs are deposited in the water to the day when,
slowly and almost painfully, the pupa emerges from
the river into the open air.   Then, when from out of

the said pupa the insect appears in its semi-complete state, it has to undergo yet another change before it becomes the perfect Mayfly.

After having divested itself of its pupal covering, it flits gently to some branch or reed, whereon it rests while waiting for the final metamorphosis. It is supposed, and I think reasonably, that this is effected by the action of the sun and air. Presumably the duration of this operation depends largely on the condition of the atmosphere. After a time a further skin splits open, and the Mayfly comes forth fully developed, in all its perfection, to join the dancing throng of insects which perform their wonderful gyrations over the waters. After so protracted a process of creation, it might be imagined that the Mayfly would be granted a fair lease of life, supposing it to escape the onslaught of anglers, birds, and fish ; but, as has been said, a very brief existence is its lot. It lives but a few hours. The Rev. J. Wood states, in ' Insects at Home,' that ' the life of the Mayfly in its normal condition is supposed never to exceed the limits of a single day.' It must be remembered that this insect does not possess a mouth, or only a very imperfect one, and therefore cannot eat : a fact which lends additional mystery to its existence. While it lives it loves : that seems the principal reason for its

D 2

creation. But though it may survive for but a brief
period, nevertheless its memory is always held dear
by all fishermen.

During the last few years considerable changes
have taken place with regard to the size and tying of
artificial Mayflies, and in my humble judgment much
advance has been made in this direction. It was not

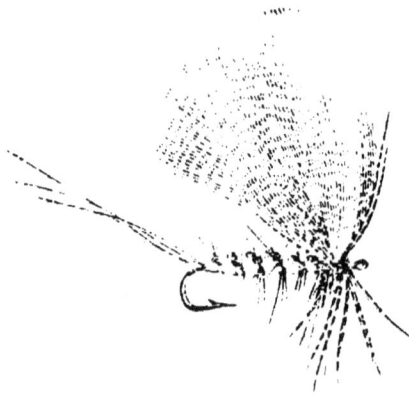

FIG. 1.—Mayfly, old pattern.

long ago that Mayflies were usually tied on big hooks
with correspondingly large wings, and became almost
unwieldy in consequence. Now-a-days the majority
of fishermen have them made much smaller, and find
that by this means they obtain better results than
formerly. The size of an artificial Mayfly such as

was generally used fifteen or twenty years ago and
that of one tied in these days is here shown.

It will be evident that under any conditions it
must be easier to cast the smaller rather than the
larger fly, and should an adverse breeze be blowing
still greater comfort to the fisherman ensues. There
may be occasions, such as very wild, rough days when
a really large Mayfly becomes valuable; but for the

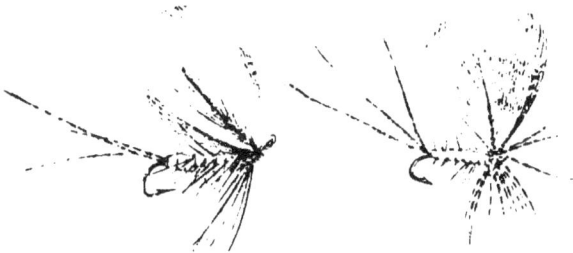

FIG. 2. Mayflies, new pattern.

class of weather which on the average is doled out to
the angler during the Mayfly season, medium or small
sized flies generally answer all purposes.

The oil which is now nearly always carried by
a fisherman is of the greatest assistance when using
the artificial insect. There is even on a small Mayfly
such an amount of body that, no matter of what
material it may be made, it must get much soaked
and sinks accordingly. Here the oil comes to the

rescue nobly, and by giving body and hackles a good touch over, a very few turns in the air for drying purposes are necessary.

Mayflies are tied in so many ways, and so many fancies abound as to the proper colours for wings and bodies, that I only dare suggest the two following patterns as those which I have found thoroughly effective on several rivers. They are :—

1. Body of dubbing, with red hackle ; wild-duck wing and mallard tail.

2. Body of straw, with red silk whipping round it ; red hackle ; wild-duck wing and mallard tail.

Another kind of Mayfly—the sunk one of which a presentment is given on p. 39—is tied something after the following plan :—

3. Straw body ; red silk whipping round it, and a good amount of bustard hackle. This last fly is without wings.

This sunk Mayfly is useful on those days when coarse weather is about. I have frequently done well with it, usually when fishing down-stream, and I think a few should always be placed in the fly-book.

The Mayfly, as will be readily understood, is generally fished 'dry' over a rising fish, and when a trout is in the mood it is probable he will be 'moved.'

But even when fish are willing and greedy, the number of rises over and above kills is nearly always remarkable.

Trout are very apt to come up and look at the fly, and then, from some cause or other—possibly a temporary form of indigestion caused by too much Mayfly banquet—hit it with their noses and retire

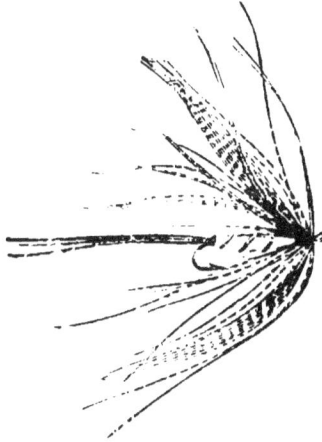

FIG. 3.—Sunk Mayfly.

swiftly. The angler very probably thinks he has had a rise, and strikes naturally without result. As a matter of fact in a case such as this the fish never opened his mouth at all, but rose merely to 'prospect.'

I have known fishermen who apparently did not deem it worth their while to continue casting if fish

were not seen to be rising or 'standing' near the top
of the water. This in my opinion is a mistake, for I
have very frequently killed trout in the Mayfly time
by throwing over likely spots when neither fish nor
natural fly could be observed. In fact, in any river
where much weed exists, the openings between the
flags, though only a few inches of water may be
visible, should always be tried. A cast which places
the fly so that it seems to drop into the river from off
a reed, as a natural insect would do, is one of the
most killing proceedings possible.

When the fly is about, I would say 'fish every
inch of water.' Trout, however large and respectable,
break through all established rules of existence when
their Mayfly feast is placed before them. They are,
so to speak, 'all over the place,' and will rise 'all
anyhow.' I know of no more nervous or exciting
moment than that when a big trout makes a rush
across a still shallow after your Mayfly. He comes
at it exactly like a torpedo boat, leaving behind him
a wave showing his track. Many things happen
under circumstances such as these. Often from over-
anxiety one strikes too soon, and the disappointed
fish sails sulkily back again, possibly bearing with
him a tender recollection in the shape of a prick from
the hook. From the same cause, if the fish *be*

hooked, perchance too violent a strike is made, and
the line comes back in the fisherman's face *minus* fly
and some inches of cast.

But should all go right, then what a glorious
moment it is ! The capture of a great trout under
such conditions after having waited heart-leaping
during the furious pursuit—is one of the happiest
events of a fishing career.

Should it be a good Mayfly season it is perfectly
marvellous in what multitudes that insect appears.
On the Kennet or Test or Itchen, or even on the
Derbyshire Wye, the very air is thick with them,
while the rivers are covered by their bodies, dead or
alive. Where the Great Western Railway runs close
by the Kennet I have seen the inside of a carriage
absolutely teeming with them, somewhat to the
consternation of a friend of mine who was not well
acquainted with the habits of this insect. When the
fly is thickly ' up ' it is almost comic to watch the efforts
of the small birds to swallow yet one more. They sit
on branches, and literally gasp, so full are they of
the succulent food. I am bound to say they do their
little best in this direction, instinct doubtless teach-
ing them that but a short ten days can be counted
on wherein to revel in their rare dainty. In fine,
brief though it be, the Mayfly season is one which,

coming as it does when the country is in the first and best bloom of its beauty, must always appeal strongly to the feelings of any fisherman.

Obviously these remarks as to fly fishing for trout do not apply to the procedure adopted when lake fishing.   Nay, as has been shown, Mayfly fishing is to a certain extent different ; so is the use of the stone-fly, and so is fishing for the big trout of the Thames. Everything hitherto suggested has had relation to the ordinary methods of fly fishing in English rivers.

CHAPTER II

LAKE FISHING

I can only deal somewhat generally with this branch of trout fishing, for it is not as if it were pursued with artificial flies tied after the likeness of natural insects, in which case much more might be said on the subject; but as lake fishing is usually conducted by means of largish flies, utterly unlike any living ones, with perhaps three on a cast, it is clear that not many suggestions can be advanced as to this section of our subject.

A good deal must depend on what are the characteristics of the lake. If it is small, it is probable that there may not be a boat on it, in which case it must be fished from the shore. If it be large, then a boat and a boatman must be secured if much sport is to be obtained. It does not greatly matter whether the lake be a real 'loch' in Scotland, surrounded by mountains, fringed by willow, birch, and rowan trees; or merely a reservoir appertaining to some great

English manufacturing town, with a huge and hideous dam-head and obtrusive sluices visible. The process of catching trout in both is very similar—even the flies used are much the same.

But let us suppose that it is some small lonely loch in the Highlands of Scotland which has to be fished. If there is no boat, then bank fishing it must be. First of all, what flies are to be used? Well, I confess I do not know the specific names of *any* loch flies, but to describe those which may be taken as stock ones—useful almost anywhere—I venture to recommend the following :—

1. Teal wing, reddish wool body, with gold twist and red hackle.

2. Teal wing, black body, with silver twist, black hackle. (Either of the above may be dressed with mallard or woodcock wing.)

3. Red wool body, with silver twist, red hackle. Wing, a white-tipped feather from a wild-drake's wing. This last is a useful fly on almost any lake in any part of the country.

These tyings are taken from Mr. Francis's book on angling, and I believe them to be standard flies which will generally kill trout. At any rate, I myself have usually found them efficacious.

There are, needless to add, swarms upon swarms

of flies recommended as being good for lake fishing.
Many of them are excellent, but they are mostly un-
named and unnamable, so should the reader wish to
procure any, he can only be referred to the fishing-
tackle shops. It must be added that several flies
used as a rule only on rivers often kill well on lakes.
Flies such as the gray-dun, red-spinner, March-brown,
and the red and black palmers (these two last especially), have often proved their value in Highland loch
and English lake.

As regards size, it is not an easy task to advise ;
but flies tied on hooks of sizes from 3 to 8, as
shown on the scale on page 106, will as a rule be
found most useful. In a gale of wind, it is impossible
to say what sized hook may not be used with advantage ! A small grilse fly certainly ; and it is a fact
that when a strong breeze thoroughly gets hold of
a Scotch loch, especially if it be in the far north,
and one perched up among the mountains, then the
big fish come to the surface and feed, while the little
ones are conspicuously absent. This applies almost
equally to any piece of water in England or elsewhere.

But to get back to our loch. As it is not a wild
day, but there is a nice light breeze flitting about on
the surface of the water, it will not be necessary to
select too large a size of fly : so two, one with a reddish

body and the other with a black one, both possessing quiet brownish wings, are chosen and attached. In passing, it may be said that an eleven-foot or twelve-foot rod is quite long enough for this work. If a larger and deeper lake has to be fished, where wading must be resorted to, and the likely spots lie further out, then a double-handed rod may with advantage be used; but in the present case it would be merely adding to the fisherman's fatigue, and could not be of any practical advantage.

The angler should begin to cast carefully where any rocks or stones break the surface of the lake; he should not neglect the edges of any rushes or weeds, and indeed he must search any water that may seem at all broken or ripply, provided there be sufficient depth to hold a decent trout; but he should always remember that in lake fishing he must keep on casting, whether he see fish rising or not; and he must also learn to work his fly slightly when on the water. The rougher the lake the more movement may be given to the flies. Fish come from out of the depths to feed on the edge of the shallows, so those the fisherman must carefully try. Of course, wherever the wind, touching the water, causes a ripple, there is a chance of a fish, and a thoroughly ruffled lake affords the most promising chance of obtaining a full basket.

Supposing that the lake, or reservoir, or loch, is one of sufficient size to allow of a boat being used, the first matter to be attended to is that of procuring the services of a boatman who knows thoroughly where the fish are in the habit of feeding, and who also can be depended on to manage a boat properly. This is not so easy as it sounds. Many a man can row a boat well enough under ordinary circumstances, who makes a mess of the business when out with a fisherman.

The boat should be taken to the 'up' wind end of the lake, and should then be allowed to drift very slowly, stern first, over the most likely spots, the boatman meanwhile pulling very gently against the wind, taking care the craft does not move too quickly. The fisherman stands or sits in the stern, and casts therefrom. Should the oarsman be inexperienced or careless, he will allow the boat to drift too rapidly, and will by this means prevent the water from being properly and carefully fished ; or he may splash unduly with his oars, and cause unnecessary and unwelcome disturbance, thereby scaring away the fish. A man who knows his business well will let the boat float almost imperceptibly in the required direction, guiding it so that all likely ripples, the favourite feeding-grounds, every inch of water wherein it is probable a trout may be, is thoroughly exploited. When a trout is hooked, the boatman should at once row hard a stroke or two *up* wind, away from the fish, and then ' hold ' the boat, so as not to disturb any fresh water.

There is no doubt that a really capable gillie, or boatman, is instrumental in adding largely to the success and pleasure of a day's lake fishing. Any one fly fishing on a lake should carefully bear in mind the following points :—

1. That the more the water is ruffled the more likely it is that the larger trout will rise ; and that the size of the flies should, as a general rule, be varied in accordance with the extent to which the water is disturbed by the breeze.

2. That every likely portion of the lake, as far as possible, should be fished, whether trout be rising or not.

3. That a certain amount of work, or movement, should be imparted to the flies, in contradistinction to the necessity of allowing a fly to float absolutely quietly when fishing ' dry ' on a river.

4. That in a lake, the best places to fish are the edges of the shallows where they abut on the deep water, and round any rushes or weed-beds. The rims of any lumps of foam which may have collected together are always worth having a fly cast into them.

Whatever causes a break in the surface of the lake is of value, whether it be a point of rock, fallen branch, or anything else which may induce a swirl in the water, or be the means of collecting together food of some description, thereby attracting the trout to its vicinity.

These somewhat disjointed suggestions on the subject of lake fishing with a fly are here offered, not as touching on the whole business of an angler when dealing with lake trout, but as having reference to most of the principal matters in connection with this branch of angling. The fisherman must learn much for himself which cannot be imparted by writing.

The secret of how to 'strike' properly when loch fishing is in itself a mystery only to be solved by experience; and how to describe striking at any time, whether when fishing up-stream with a dry fly or on a Highland loch, is a matter which I confess my inability to attempt. 'Striking' consists of tightening the line sufficiently to make the hook fasten in the fish's mouth as soon as ever the fisherman conceives the said hook to have entered thereinto. Some have described this act as being accomplished by a 'turn of the wrist.' I frankly say I cannot conceive how such a proceeding can achieve the desired object. As far as I can judge, all that could

E

happen should this manœuvre be executed would be
that the point of the rod would drop slightly and
the line thereby be slackened, the very thing to be
avoided at such a juncture ! The object of the 'strike'
is, as has been pointed out, to drive the hook in over
the barb into the fish's mouth, before he has had time
to discover that the fly or bait, whatever it may be,
possesses an unnatural and disagreeable foreign body,
and consequently endeavours to spit it out. The
strike can be effected only, as far as I can see, by raising
the point of the rod at exactly the right moment, not
so violently as either to drag the hook out of the fish's
mouth, or to break the casting-line. I have often
watched skilful fishermen rise, strike, and hook fish,
and I have always seen the rod go *straight* back,
quickly but gently, in the direction of the angler's
shoulder. Any lateral motion such as that suggested
by the expression 'turn of the wrist' has been most
carefully avoided.

With very fine tackle a fly is frequently struck off.
No man, however great an adept he may be, can help
this happening at times. An extra-cramped position,
or a very long line which has a slight drag on it, will
make such an accident very possible. If the fisherman be not quite fit and well, if his nerves are
slightly upset, he is apt to strike somewhat more

hastily and hard than is usual or desirable ; and a smash will then most probably ensue.

When trout are large it seems requisite to let a longer time elapse before the strike is made, whether in lake or river, than when they are small ; and I cannot help thinking that as a rule fishermen err on the side of striking too rapidly. Certainly this appears to me to be the case when using the artificial Mayfly. When the river fished is a rapid, rough stream, where trout do not run to a large size, then the quickest possible strike must be made, because it is probable that in such water the rise is not noticed till it is half over ; and also because, the current being swift, the whole operation has to be correspondingly nimble in its execution.

But to return to lake fishing. What I have hitherto said in relation to this subject has been exclusively in connection with fly fishing. There are other methods of killing trout in lakes ; indeed nearly all the monsters of which any records exist have been caught by spinning, either with a phantom or a natural minnow. In Ireland, on the Westmeath lakes, 'dapping,' or using the blow line with the natural Mayfly, accounts yearly for the great majority of the big trout captured in those charming waters.

Mention of Ireland suggests the introduction here

of some notes on these same Westmeath lakes, which
have been supplied to me by the courtesy of Mr.
W. J. Thomas, Honorary Secretary of the Angling
Club at Mullingar. Mr. Thomas says :—

'The Westmeath lakes are generally recognised as
Loughs Derravaragh, Owel and Ennel. The first
named is situated about five statute miles from the
county town of Mullingar, and the others two statute
miles equidistant from the same place. All three
lakes teem with trout, having been carefully preserved
during the past eleven years by the Westmeath Lakes
Fish Preservation Society ; supported by voluntary
contributions and a grant of 25*l.* from the Limerick
Board of Conservators for assisting in preserving the
salmon in the rivers Inay and Brosna.

'The season for trout opens on March 1st and
closes on October 1st.

'When the season opens, the usual practice with
the professional fishermen and boatmen is to troll
natural bait on rods or hand-lines, particularly on
Lough Owel. The baits are small trout, perch,
gudgeon, loach and minnow, and by this means the
larger fish ranging from 5 lbs. to 12 lbs. are taken.

'The legitimate anglers use artificial flies of various
patterns, the best being those tied by local fishermen,
comprising different shades of olives, cinnamons,

clarets, black-hackles, orange bodies, hare's lugs, &c.

'They also use for trolling, artificial baits such as Devon minnows, spoons, phantoms, &c.

'When the natural fly, commonly called the green drake, begins to rise in May, a regular fishing carnival sets in. At this period a light handy rod, usually made of two joints of well-seasoned ash, with a top joint of lancewood, is generally used, sixteen feet being the average length. The reel should be capable of holding fifty to eighty yards of good silk line, on to which is whipped some twelve to fifteen yards of soft, light, and strong silk line called a blow line, suitable to be carried a fair distance from the bait even by a light breeze.

'To this blow line is appended about four feet of fine drawn gut and a small well-tempered hook. The natural flies are collected and kept for the day in small boxes or baskets. Two of these flies are impaled on a hook, and the blow line, by the action of the breeze, carries the flies sufficiently far from the boat, where they float or "dap" or "dip" on the surface of the water.

'It should be mentioned that in this particular mode of fishing the boat is allowed to drift before the wind.

'On a favourable day twenty fish for a boat of three rods would be considered satisfactory, the trout ranging in size from 1 lb. to 5 lbs. each.

'I give three special days in my own experience. On Lake Owel some few years ago my wife and myself had 27 trout weighing 55 lbs. ; and on another occasion I had 8 trout to my own rod weighing 39½ lbs., and on Lake Ennel or Belvedere my wife and I had 14 trout weighing 30 lbs. The largest I have ever killed was one 12 lbs. in weight, "dapping" on Lake Owel, but I have seen them taken at 14 and 16 lbs. I have frequently caught them 4, 5, 6, and 7 lbs. each.

'The largest trout ever caught was by rod and line and spoon bait in Lake Ennel near the shore of Belvedere demesne, by Mr. Willie Meares of Mullingar, on July 15, 1894. It weighed 26 lbs. 2 oz. I had the fish set up. It is lent by me to the Congested Districts Board of Ireland, and is now (June 1897) on exhibition in London. The trout of the Westmeath lakes are of excellent flavour, and during the open season are in the pink of condition.'

This description of Irish lake fishing embraces all the methods employed in that part of the United Kingdom for the capture of trout. It will be noticed that every species of bait is mentioned, artificial flies, natural flies, live bait, and artificial bait ; and it will also be remarked that the trout run to a very large size.

Returning for a moment to the subject of lake fishing in general, should the fisherman be spinning, he must be guided to a large extent, as regards the size and nature of his bait, either by the advice of any local fisherman he may encounter, or by his own observation of the characteristics and requirements of the lake he is fishing. Should it happen that he is not previously acquainted with it, he should make every endeavour to find out from some one living in the neighbourhood whether the trout prefer spoon, phantom, or natural minnow, live or artificial bait. In lakes, trout vary in their tastes nearly as much as they do in rivers, and it requires careful notice and considerable practical knowledge to become a successful lake fisherman.

CHAPTER III

WORM FISHING AND SPINNING FOR TROUT

IN these days, when the attention of fishermen seems to be so firmly fixed on fly fishing as almost totally to exclude most of the other branches of the sport, I feel somewhat nervous when I allude to the use of the worm as a means of catching trout.

This, however, is but a species of false modesty, and very false at that; for I know few things pleasanter in the way of fishing than to wander amongst Highland hills, wherever some wee burn may run, cairngorm-coloured, making miniature cascades over the boulders, or gliding quietly over pools which from the intensity of their colour seem trebly deep, equipped with a light, somewhat stiff rod, some Stewart tackle, and a bag of well-scoured brandling worms.

True, the trout may not run large, in fact they are sure to be small, and the stream may be only of a size which would put fly fishing almost out of the question. But what of that? Are you not, as far as your own personal ideas are concerned, monarch of all you survey? Probably you do not own an inch of all the glorious country around you, but you have leave to fish this particular burn, and that fills your cup of happiness. It is equally probable that, in giving you permission, the owner of the property looks upon you as a species of harmless lunatic, for the smallness of the stream and of the fish in it can be attractive only to a keen trout fisherman. But what does that matter to you? As you stroll along the burnside, through the rough fields which trend upwards towards the hills forming the fringe of the moor—fields full of queer tussocky patches, half rush, half grass—you feel, right to the very marrow of your bones, the magnetic influence of the Highland atmosphere. To my thinking there is nothing in the world so bracing and inspiriting, or at the same time so nerve-comforting, as this same Scotch air ; and so be thankful for it. As you wander on before beginning to fish, you become aware of the clear whistle of the golden plover and the call of the peewit. The pert wagtail flirts his tail, the dipper winks his eye at you, and then quietly vanishes. In the distance you can both see and hear grouse ; in short, on every hand there is infinite

variety for appreciative and comprehending eyes to
observe.

But now you have reached the spot where you
should start fishing, so put your rod and tackle
together and get to work.   Hereabouts the heather
begins to spread down to the side of the tiny stream,
and thus adds to the physical comfort of the fisherman ; for he can fish all the water while walking or
kneeling on this softest of carpets.

Now, what sort of hook will you have fixed to
your cast, which should not be more than some four
feet long ? A single hook, or a Stewart tackle ?
The sizes shown in the accompanying diagrams
are those which, as a rule, are most advisable.
In the case of the single hook you are less liable
to be 'hung up' in stones, or bank edges, or other
obstructions, supposing the stream you are fishing to
be a small one and much intersected with boulders,
rocks, and other fishermen's worries ; but, on the
other hand, with the single hook you are much more
liable to miss your fish when striking.   With Stewart
tackle, either double or treble hooked, you do not
lose so many trout as when using only a single hook,
but naturally you get more often caught up in the
hundred and one things which seem to be placed in a
river principally for the purpose of testing the measure

of the angler's temper.   But taking it all round, I do
not think that any one who has thoroughly tested
the merits of both kinds of hooks would hesitate in
selecting the Stewart tackle as the best to be used
under circumstances such as those with which we are
now personally concerned.

So put on the Stewart tackle, double-hooked (for
the burn is small), without any shot on the cast, as the

FIG. 4.—Single hook.

weight of the hook and its accompaniments suffices.
And now as to the bait.   You have in your basket a
box of tough, not too large, brandling worms, which
have been for a day or two in fresh moss, and have
thus had time to clean and harden themselves.   Also
you have, attached to a waistcoat button, a small
canvas or brown holland bag with a thin wire run round
the lip of its mouth, so as to keep it open sufficiently

to enable you to put in your finger and thumb when you want a fresh worm. This is wise, for it wastes endless time to have to get the worm-box out of the basket and then put it back again on every occasion when a change or replacement of bait is required. By means of the auxiliary bag much rapidity is added to the incessant operation of putting on worms, and the box in the basket acts as the necessary reserve. So place a sufficient supply of worms in the

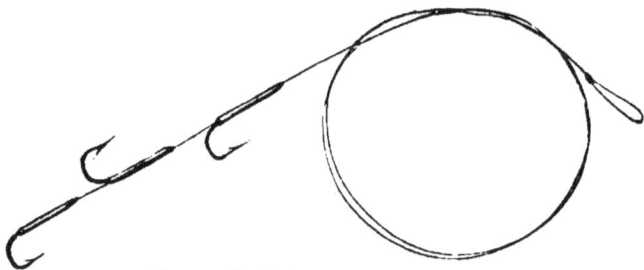

FIG. 5.   Treble-hooked Stewart tackle.

bag and get to work. Try every tiny run, every inch of water which will cover even the smallest trout. When the worm is the bait all ordinary rules are held in abeyance. You must *swing*, not *throw* the hook. If you cast in the usual fly-fishing manner, it will constantly happen that the worm jerks off. As your rod should be somewhat stiff, this 'swinging' is the more easily managed. You are of course fishing *up-stream*, and whenever you see the slightest check to

your line take place, strike lightly. It may be a trout
and it may not ; but in any case strike. If it prove to be
a trout, it will probably not take you long to get it
out : in fact, you will as likely as not pull it straight on
to the bank, and very possibly it will not weigh more
than an eighth of a pound, if as much. But, neverthe
less, you will at the moment be as pleased at its capture as if it were a fish of five times the size, caught
in more southern waters.

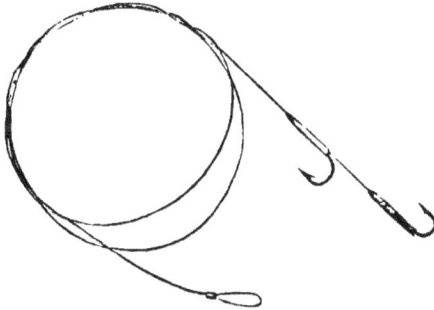

FIG. 6. Double-hooked Stewart tackle.

And thus you go on fishing, always up-stream, till
you arrive at the inner recesses of the moor, where
the scenery grows wilder, where you feel yourself in
still closer communion with Nature, and more under
the spell of her beneficent influences. By the time
you begin to retrace your steps and fish in a casual
and satisfied fashion down-stream towards home—for
have you not caught between four and five dozen

troutlets?—you may have undergone some small
adventures; such as having, round the corner of a
'knowe,' chanced suddenly upon a shepherd's hut,
from the vicinity of which bounced out a couple of
wild sheep-dogs apparently desirous of immediately
devouring you. By means of hurriedly acquired
stones you will have routed these foes and escaped
as quickly as your dignity permitted from their
immediate neighbourhood. If the month be August
or September, you may have seen the pursuit of
the grouse by the owner of the moor and his friends;
you may have slipped from a stone into the water
and got wet through, or many other things may have
occurred.

But whatever may have happened, if you have
had any measure of success, I am sure that you
will agree with me that, through the medium of
the much-despised worm, you have passed a thoroughly delightful day amongst glorious scenery, in
health-compelling air; and that though the size of
your victims is not large, nevertheless you have
enjoyed yourself hugely, with much benefit to both
body and mind.

But worm fishing for trout is not confined solely
to the smaller streams. A day's Stewart tackling in
a Scotch river such as the Don or the Deveron,

during the summer months, when the sun is abroad and the shallows visible, is not easily forgotten.

You must wade, and your rod must be fairly stiff — one of about eleven feet long will serve. In all other particulars your tackle should be the same as that which you would use when fishing a wee burn or beck, always premising that you have a reel holding sufficient line to compete with the fighting qualities of a bold, strong, pound and a half fish. For in rivers such as these the trout run to this size and larger.

In direct contradistinction to the usually accepted doctrine which lays down that clouds are a valuable help towards a good day's sport, the sun is hailed by you with approbation ; for it is in the shallower water wherein the trout lie basking, enjoying the warmth, that on such occasions you will to a large extent rely for the best part of your day's sport. Your proceedings must be of much the same nature as when 'worming' a smaller river ; except that in all probability you will kill many of your fish close to, or on, the aforementioned shallows.

In wading up the clear runs—for you must fish up-stream—you should move as gently as you can, so as to cause as small a disturbance of the water as may be contrived. When you have hooked a fish do not

leave the river to get it out, but have a short handy
landing net, so that you can secure it in mid-stream,
thereby saving time and preventing unnecessary alarm
to the adjacent fish.   It is wonderful how determined
trout are, when fairly on the feed. They will then sometimes seize the worm almost at the fisherman's feet;
and should a lively fish on such an occasion take it
into his head to charge the angler, it is more than
likely that a very pretty complication and ultimate
catastrophe may occur.

Large baskets are made by Stewart tackling in
bright weather, and it is a branch of trout fishing
which necessitates to a considerable degree a keen
sight, a quick and light hand, and much practical
knowledge of the habits and customs of trout under
the weather conditions in which it is usually pursued.

But there is yet another way in which the worm
can be utilised for trout fishing, and I have known
this fashion to yield excellent results and to afford
capital amusement.   Wherever this bait is permitted
it is feasible.   I mean worm fishing with a single
small hook and one or two shot on the casting-line.
The strength, size, and volume of the river must
decide how much the line will require to be
weighted.   The hook is as a rule of the size
numbered 1 in fig. 7. p. 106.

The general plan of operations is somewhat as follows :—

The fisherman, if he knows the water well, at once marks the most likely place for trout, having due regard to the state of the river : and instead of wasting unnecessary time will begin at once to fish for them. Across, up, or down stream is the order when using this kind of tackle, and personally I have seen more execution done when the hook is being worked just *opposite* the angler than when in any other position. The edges of runs where the water will move the worm steadily ; deep holes ; almost any place into which the hook can be dropped without being ' hung up ; ' these are the spots to try ; and on streams, such as the upper waters of the Till, I have seen many and large trout killed by this method, especially after a slight 'spate,' when the water was 'clearing off.'

It is very pretty work, moreover, and work wherein one who is acquainted with the inner mysteries of this branch of the gentle art can demonstrate to the full the fact, that in worm as in fly fishing a professor will generally handsomely defeat a novice.

Not only in comparatively small streams is worm fishing with a single hook a deadly form of angling. In rivers like the Tay when a flood is just running

F

down and when, though big, the water is beginning to show signs of being again fishable, I have witnessed the capture of large basketsful by the method we are now discussing. Forced from their habitual haunts by the pressure of the mighty waters, the big trout take up their positions nearer the banks, well aware as they are that a 'spate' brings down much food, swept by the rise of the floods from the river sides, and that these viands are more likely to be found near the shores than in mid-stream. Thus the anglers are enabled to reach fish which under ordinary circumstances would probably be unattainable. With heavy water two or three shot will be required to sink the hook sufficiently. Well do I remember being initiated into this form of fishing by one of the best professional fishermen on the Tay, near Dunkeld, and the wonderful results which he obtained. In fact, any one who knows how to avail himself of the worm as used in this fashion can, when the opportunity arrives, kill trout when no other kind of bait would be of much service.

What shall be said about a branch of trout fishing in which it is a work of time and experience to attain to real excellence—I mean spinning? Shall explanations be attempted as to the best methods of working the natural minnow on spinning tackle?

Shall I try to explain in what manner the mighty trout of the deep and far-extending Scotch and Irish lakes are captured by the trolled 'phantom,' or how that craftiest of spinners, the Thames fisherman, obtains his hard-earned reward ? Or shall I tell how to wile the small but hard-fighting trout of the lesser streams by means of the tiny 'Devon' minnow, or brightly gleaming spoon ? No ; I shall not endeavour to do any such things, for, without unduly extending the scope of this work, I could not adequately deal with this branch of the fisherman's business.

So I will say briefly that, in spinning, you must, as regards rod and tackle, be guided by what kind of water you are about to fish, and make your preparation accordingly.

If you are going to 'spin' a small Devonshire river, then your ordinary fly rod will serve to work the little artificial minnow which you will probably employ ; whereas if you are going to try your luck on a great Scotch loch, a stiffer rod and a longer line are obviously required wherewith to 'work' your larger bait. Experience teaches these things—nothing else can effectually do so.

I have only one other remark to make on this subject ; which is, if the fisherman is about to try a river on which the use of worm or spinning tackle

is allowed as well as the more legitimate artificial fly,
let him not fail to put in his fishing-book two or three
'Spinning Alexandras,' attached to fine gut traces.
These bunches of peacock's feathers and silver tinsel,
with their triangles of small hooks, are in general
most efficacious.   By their aid I have often caught
good trout, when I have been permitted to use this
so-called ' poaching ' arrangement in water which had
proved perfectly obdurate to other angling artifices.

Whenever a river is ' running off' after a flood, the
edges of the heavy water should be searched by
the Alexandra : under bridges in the deeper parts of
the streams it works wonders, and as the ordinary fly
rod is available for this manner of fishing, it has the
great merit of convenience.   Moreover it is a pretty
sport, and one which is very likely to add a store
of good trout to the fisherman's basket when other
methods may have failed.

CHAPTER IV

RIVER PRESERVING AND RIVER POACHING

A RIVER-KEEPER who is thoroughly conversant with the habits and requirements of trout, who recognises at once the various flies as in due season they appear, and who notices what amount of weed there is in the river, how much should be cut and how much left, is indeed a treasure, and ought to be valued by his employer accordingly.

An expert in these matters will almost instinctively be aware that, to keep the river at a proper level both for fishing purposes and for the benefit of the farmers who hold riverside farms, a certain amount of weed and water plants is essential. He should consider whether the past winter and spring have been wet or dry, and calculate how much weed-cutting may judiciously be done to render the river fishable, while not removing so much as would cause the water to fall unduly and in consequence deprive the fish of their requisite shelter and food-producing haunts.   He will require also to allow for the 'heading back' by mills and eventual flow of the pent-up water.

Much attention must be given to the condition of the banks, and to the silting up of the deeper channels of the river, if it be a small one, and therefore easily dealt with.

These and many other matters provide full employment for an energetic river-keeper all the year round, and in his case constant vigilance is as necessary as it is in that of a keeper whose business is amongst the woods and fields.

Where the river is of considerable size, such as the Test or the Itchen, the fishing keeper's task is a somewhat easier one than on a stream of the calibre of the Lathkill or the Lea ; as the more vigorous flow and stronger volume of water in the former rivers take to a certain degree the management out of his hands, and he can only deal with those portions of the stream which are accessible to him.

Touching the question of poaching in connection with the subject of a river-keeper's duties, there ought not to be — in England at any rate — any serious difficulty in defeating the nefarious designs of those who wish to interfere illegally with trout fishing.

Putting on one side any idea of a river being

poisoned by lime, or of the use of explosives to
destroy the fish, against which proceedings heavy
legal penalties exist, the only other plans a poacher
can adopt are netting or setting night lines ; and if a
keeper is worth his salt, he should easily outwit any
one attempting either of these methods.

Netting can only be practised openly ; and though
a keeper cannot be on his beat night and day, constant watching will do much towards warning bad
characters away from the river. It is most necessary
for a keeper to possess a knowledge of the regular
poachers of the neighbourhood.

It is not likely that every portion of the water
is available for netting purposes, but in those places
where such a proceeding is possible, a stout, square
stake firmly driven into the river-bed, ornamented on
all sides with strong up and down turned hooks, will
tear to pieces any nets which may be used.

These stakes should reach to within some eighteen
inches of the top of the water and will interfere little
with the fishing, while they prove an almost insuper-
able bar to successful net-poaching. Night lines, on
a river carefully watched, produce such feeble results
that they are very little to be dreaded.

In a word, a simple device, such as described
above, really meets all the requirements of the case,

provided always that the keeper and his subordinates know their duties, and perform them.

But, however excellent a river-keeper may be, lessees and owners of trout-fishings may themselves do much in the direction of the prevention of poaching. They should always make friends with the farmers who are tenants of riverside holdings, and as far as possible consult their wishes with regard to the state of the water in the streams. They should get to know the labourers who work in the fields near the rivers; and if they do this they will often find themselves in possession of valuable information – probably conveyed in some roundabout fashion—respecting the doubtful characters who may inhabit the district.

Labourers who are thus treated, and occasionally —if they have helped the fisherman in any way— judiciously 'tipped,' often prove most valuable amateur watchers, and are thus of much assistance to the regular keeper. And, above all, every one who is interested in any river should take care to impress on those who fish in it that nothing goes further to promote good feeling and a spirit of friendliness between anglers and riverside tenants than the avoidance of all unnecessary damage to crops growing in fields which stretch to the river-banks. A certain knowledge of agricultural requirements is a good thing for an angler to possess. Nothing is more aggravating to a farmer than to find that a fisherman has taken a short cut across a field of strongly growing mowing-grass, leaving behind him a path deeply beaten down by his heavy wading-boots, when a détour or retracement of fifty or a hundred yards would have obviated any such destructive proceeding.

Those who have fished any South-country river, Test, Wiley, or other delightful stream, will at once recall memories of lush water-meadows, while Derbyshire men will conjure up visions of Haddon Fields just before hay-cutting time. I am sure that injuries to crops are nearly always done unthinkingly, but I cannot refrain from here stating my feelings on this matter, which I am convinced is one of considerable importance to the fishing public. It is, of course, impossible to avoid doing *some* small hurt to the grass on the bank of a river; for if one does not move along the streamside one cannot fish — that is perfectly clear. But this is not what I allude to; it is the straying away in an uncalled-for degree from the strip of ground by the river-bank which is always dedicated to the fisherman's use; that it is which I would deprecate, and on which I have thus briefly commented.

But unless some extra-malign influences exist, there ought to be no great difficulty in properly preserving

trout waters in England, or in keeping them free from
poachers.

While on the subject of river preservation it would
appear that, as a natural sequence, some brief allusion
should be made respecting the steps to be taken when
stocking a stretch of water, whether it be a lake or
a river, so as to render it available for the reception
and existence of trout.

I must at once admit that it is one thing to say
' I will' advise on this matter, and quite another
when it comes to carrying out the intention. For
streams and lakes vary so much in their nature
and trout-bearing capabilities, that to generalise
concerning them is a somewhat delicate undertaking.

The reader will find, in the part of this volume
dedicated to Trout Breeding, such full and admirable
descriptions of each and every process which is in
vogue at an establishment such as Colonel Custance's,
that not another syllable is required to elucidate
the mysteries of trout production ; and, therefore, the
initial stage, namely that of obtaining information as
to the best species of trout wherewith to stock new
waters, is at once surmounted.

But different rivers have different peculiarities ;
and it is by no means certain that fish which will

flourish and do well in one stream will thrive equally in another which may be only a few miles distant.

The various strata of the soil may be dissimilar : one river may flow through and over chalk ; the other through gravel or perhaps heavy clay : though, as a general rule, such discrepancies do not occur within a limited area. With regard to this, two rather remarkable instances have come under my own observation within the last few years. One occurred in connection with a river, and the other with a moorland reservoir.

In the former case a number of fry and yearlings, as well as a few two-year-olds, were procured from a well-known Scotch trout-breeding establishment and placed in the Derbyshire Lathkill. They arrived in excellent condition, and presented no peculiarities whatever. I should add that they were the ordinary Scotch trout, not Loch Levens, or any sort of 'fancy' fish ; and the river in which they were placed has many of the characteristics of a largish Scotch burn or Highland stream. In fact, as far as I could see, the new trout would find their surroundings—with the exception, perhaps, of insect food being rather more prevalent in the Derbyshire river than in the more Northern waters—very similar in almost every respect to those which would have environed them had a Scottish stream been their fate.

For a couple of years all went well. They rose more freely than did the home-bred trout, and were very 'cheerful' fish, easily distinguishable from their Derbyshire brethren by the different arrangement of spots on their bodies. Shortly after, however, the keeper remarked that they did not appear to 'come on' at all; but rather that they seemed to be growing only in length, while in depth and thickness little or no improvement was noticeable. He informed me of this, and I found on careful investigation that such was undoubtedly the case; with the result that I have had to give orders that every 'Scotchman,' however small, when caught, is to be killed. For it is certain that hardly any of these imported trout, after having reached a weight of a third of a pound or thereabouts, flourished or grew healthily. They seemed simply to elongate : the three-quarters of a pound and pound fish consisted of nothing but lank discoloured bodies, and huge, hideous heads. They appeared never to attain any sort of condition, and from their feeble state afforded no sport at all. It was curious to look into a pool and see a 'Scotchman' lying side by side with native Derbyshire-born trouts. The one black, feeble, and useless; the others bright-coloured, active and well-conditioned.

I cannot explain *why* this should have been so, but that it *was* the case I can confidently aver.

In the other instance to which I have referred, namely that of a small reservoir, the circumstances are these.

Fifteen or twenty years ago, the said reservoir was constructed in the heart of the moors in Derbyshire. It is on high land, about 900 feet above sea level, and is fed, not only by the drainage from the moors, but by a lovely little burn, which from its appearance would seem to have been made by Providence specially for the purpose of containing many merry troutlets.

The reservoir was stocked with trout carefully selected on account of their suitability for existence in water such as this. What has happened? Why, that every fish has disappeared.

Fresh consignments have from time to time been added, but always with the same result. In fact, there is no fishing to be had in this reservoir, or in one or two others adjacent, as they also have developed the same unsatisfactory peculiarities. Moreover, the very burn refuses to sustain the trout which were placed therein, and not a fish can now be found in it.

Poaching has not been the cause of this catastrophe. The moors are too carefully watched for that. And though analysis has failed to discover the reason, the only conclusion I can arrive at, as regards the burn at least, is, that it possesses some strange peaty poison which destroys every trout. As to the reservoir, I must conclude that either from the same cause, or from the fact that, in common with other similar constructions, its bottom is made of 'puddled' material, it is useless to hope that any trout fishing will ever be possible in that piece of water. It is probable that the shortness of feed consequent upon the manner in which the reservoir was of necessity made, and the materials used, has been the main cause of this disastrous result; as I conclude that, notwithstanding the fact that a number of trees and shrubs, such as willow, alder, and rowan trees, have been planted round the edge of the water, still the bottom of the reservoir cannot produce any food worth mentioning. But this cannot be considered *quite* a satisfactory explanation of the difficulty; for it is certain that within a comparatively short distance there are similar sheets of water which hold quantities of trout.

I have mentioned these two cases in order to show that, even when the would-be trout preserver has settled whence he will procure the fish wherewith to stock his river or lake, his difficulties are by no means over.

Doubtless the ordinary *Salmo fario* will exist and do fairly well in most streams which present, even to an uninitiated eye, a 'fishy' appearance; though I have quoted one glaring exception to this rule. But whether they will do more than moderately well is another question altogether.

On this point—namely, what should be done to a river which it is intended should be made into a trout-bearing stream—much might be said; in fact, a great deal more than is convenient for this book. But I may venture on a few suggestions.

Supposing that the river to be taken in hand is one possessing ordinary trout-carrying capabilities — by which I mean that it has a fair share of gravelly runs as well as deep water, and is not one of which the bed is composed entirely of deep muddy deposit then the affair is comparatively easy. It will probably be advisable to remove a certain amount of mud from the river bed in various places, so as to provide deep lengths and holes in which the larger fish may flourish. This, however, should be done with due consideration ; for, when deepening channels, much bottom-feed is unavoidably destroyed, as well as the roots of many weeds, water plants, &c., these latter being essential to the welfare of the fish.

Where there are long, still pieces of water, a few half-dams may, very likely, with advantage, be inserted. These are constructed of two or three upright posts driven into the bottom of the river, to which planks are attached on the up-stream side, the top board being movable, as it may be necessary to shift it should there come a heavy flood.

The object of these half-dams is slightly to divert the current of the river from its usual placid, level flow, so as to create more movement, and some broken water.

For this reason half-dams are generally inserted in pairs, on opposite sides of the stream, one being about ten yards higher up the river than the other. They should project a sufficient distance from the bank to create by their joint presence not only a swirling stream, but two backwaters, one behind either dam, which are always sure 'holds' for trout, and those usually large ones.

To be thoroughly satisfactory as a trouting stream, a river should be possessed of a fair share of gravelly lengths on which the trout can spawn, as otherwise constant re-stocking will be required. Another matter which must be carefully attended to is the quantity of weed that should be removed from the water. I have alluded to this elsewhere, and will, therefore, only repeat that, speaking generally, it is better to err on the side of leaving too much than to remove an excessive quantity of the various water plants. In the one case you render fishing more difficult, but you preserve your trout and keep up a good average of weight; whereas by over-weed-cutting you render fishing, or rather casting, easier, but at once diminish seriously the weight and condition of the fish.

In very prolific rivers there is always a danger that the number of the trout may produce a corresponding decrease in their size; and as to this the owner must decide whether he wishes to have a more limited quantity of fish, but those of greater average weight, or to have his stream stocked with many, but smaller trout. On most waters I cannot help thinking that a happy medium as regards this point is the most satisfactory solution.

The times of the year in which trout should be placed in the rivers to be stocked are as follows : fry during February, March, or April ; yearlings about January or February ; two-year-olds during the winter.

G

I do not propose to enter upon a disquisition respecting the merits of one kind of trout over another for river-stocking purposes, beyond saying that for all general sporting qualifications the ordinary trout (*Salmo fario*) seems to meet every requirement. One curious thing which has been noted a vast number of times is, that imported trout, should they possess any small differences of markings from those fish which are 'home-bred' ones— natives of the river—after a short time seem to adapt themselves to a large degree to their new surroundings, even to the almost exact scheme of colour and spots appertaining to the original inhabitants of the stream.

Trout take the prevailing hue of the water and the immediate local characteristics in a remarkable degree. Where the river runs over a muddy bottom, there they will be dark of appearance, and somewhat muddy of taste. Where the stream flows over a clean gravelly soil, there they will be bright in colour, and far pleasanter to eat. I know of a certain length of water in a Derbyshire river where for about 150 yards the bottom is all gravel and sand. There the trout are quite brilliant in their colouring, orange as to their bellies and sides. Immediately above it comes a small piece of 'flat' water. There the fish are utterly uninteresting both as to hue and taste ; above that, again, a fair-sized waterfall broadens the river into a beautiful pool with a steady stream flowing through the centre. In this the trout recover their 'clean' colour, but seem to acquire the dark greenish hue here predominant : a tint derived from the overhanging trees at one end of the pool ; from the lichen-covered wall which forms its boundary on one side ; and from the green meadow which slopes down to it on the other.

As these differences all occur within a distance of about 500 yards, it is abundantly evident that the appearance and colours of trout are greatly influenced by the nature of the water they inhabit, and the surroundings with which they are encompassed.

Should a river be one with but few runs or 'sharps,' it is a good plan to throw in some big blocks of stone where it is stillest and slowest. These serve two purposes. The first is, they break somewhat the dull current of the water ; and the second is they at once afford hiding places and 'stances' for the fish. It is almost a certainty to find a good trout established in the back water formed by a thrown-in rock, within a very short time of the stone having been put there. Moreover, water weeds and aquatic plants at once begin to appear around

G 2

the newly imported obstacles, and thereby add to the food-producing capability of the river.

In the case of a lake, whether small or large, much the same procedure is necessary when making it into a trout preserve. Naturally any sheet of water which has a brisk stream running through it, with a good inflow and outflow, has, from a trout-carrying point of view, a considerable advantage over one which is not so favoured. For trout, whether in lake or river require clean, running water to keep them healthy and well conditioned.

The difference between a trout taken out of a muddy lake and one from a clear stream is very marked. The first is dingy and dark, with a nasty dull taste when cooked; the second is bright in colour, with a sweet ' clean ' flavour. But given that the lake to be stocked is one with fair natural advantages, the initial step to be taken is to discover whether there be any pike in it. If there are, all possible means should at once be employed to destroy every one of those trout-destroying pests. Personally I should get rid of every other kind of coarse fish as well, except perhaps a few perch, which are valuable as being excellent eating.

The pike having being destroyed, the next thing to be considered is what kind of trout should be encouraged. And here I can only say that the proprietor of the water must use his own judgment as to whether common trout (*Salmo fario*) : Loch Leven trout (*Salmo Levenensis*), Fontinalis (*Salmo fontinalis*), or Rainbow trout (*Salmo irideus*) are most suitable. Of course if there are any other adjacent trout lakes near, the owner had better enquire what species do best in them. But if he has to make his own decision he may find himself in a difficulty.

Probably the best thing he can do is to order the common trout : as, though the Loch Levens are beautiful fish to look at and admirable on the table, they are very particular as to where they live, and do not flourish in many lakes. I have seen them do well in Yorkshire and Dorsetshire, as well as in their own home.

Rainbow trout and *Fontinalis* I know very little about ; but I have been told by several of the principal trout-breeders that the latter do well in both English lakes and rivers, are bold free risers, and extremely game. As yet, however, I cannot quite make out whether they agree with the *Fario*, should they happen to be in the same water together, or whether they have rather a tendency to oust the native fish.

But whatever species of trout are placed in river

or lake, the principal matters to be remembered in connection with their welfare are, that the water should be kept as pure and uncontaminated as possible ; that 'hide' and food for the fish are absolutely essential, so that much care must be taken to avoid over-thinning when weed-cutting ; and that a good fishing keeper—one who is constant in his attention to the water, and to the needs and necessities of his finny dependents—is invaluable.

If these matters are attended to, then, given ordinary good fortune, there should be no great difficulty in persuading trout to live and prosper in most rivers or lakes in Great Britain.

CHAPTER V

FISHING GENERALITIES

CHARLES KINGSLEY, in that most delightful of all fishing tales, 'Chalk Stream Studies,' at one juncture admonishes his pupil to 'take off that absurd black chimney-pot, crawl up on three legs, and when you are in position, kneel down.' It is far easier nowadays to procure proper and suitable requisites for fishing, whether as regards clothing or tackle, than it was forty years ago when Canon Kingsley wrote. Presumably in this instance the fisherman was wearing a tall hat. Fancy trying for Test or Itchen trout in these days, clad in such a headgear ! If it were anything like a sunny day the fish would wonder, awestruck, for the fractional part of a second and then flee *en masse*.

Of course at the date when Canon Kingsley went a-fishing, trout were easy to catch compared with what they are now, at any rate in the best known English rivers ; but it will be noted that even then the fisherman was advised to make himself as inconspicuous as possible. How much more necessary is that precaution now, when trout, from being much fished over, in many cases from over weed-cutting, and from a frequent shortness of water consequent upon the present drainage system, are highly educated and more difficult to kill than ever they were before ?

This being so, it behoves any one who is desirous of obtaining his fair share of piscatorial good fortune to take care that his fishing garb is sober of hue and not of a nature to attract the attention of the trout.

Very light-coloured clothes are strictly to be avoided, and this especially as regards the cap. All fishing garments should be warm and comfortable. Formerly waders and waterproofs were not thought as indispensable as they are now ; but, as I know from painful experience, kneeling and crawling about in damp riverside grass produces rheumatic symptoms to a most disagreeable degree. Any good firm of waterproof manufacturers will supply a pair of fishing boots or fishing stockings which will come well up over the thigh, and enable the fisherman to kneel or crawl through wet grass, or wade—at any rate in most English trout streams—sufficiently deep to fish water which he could not conveniently cover from the bank, without fear of any evil results ;

while a light mackintosh is not heavy to carry in
the fishing bag or basket, and helps to make him
secure against rain. It is of no earthly use to get
wet unnecessarily, and no one kills more fish by
declining to take precautions such as these. The
fisherman must find out for himself what kind of
waders suit him best, and this can only be done by
experience. So nothing but general suggestions such
as the above can be offered on this head.

It is worth while noting that, notwithstanding
the amount of literature on the subject of angling
which has been published of recent years, no more
sound general advice as to trout fishing has ever been
set forth than that which Isaac Walton propounded
as having been given by one Mr. Thomas Barker in
his directions for fly fishing. Here it is : 'First let
your rod be light and very gentle : I take the best to
be of two pieces. And let not your line exceed
(especially for three or four lengths next to the hook)
I say, not exceed three or four hairs at the most ;
though you may fish a little stronger above, in the
upper part of your line ; but if you can attain to angle
with one hair, you shall have more rises and catch
more fish. Now you must be sure not to cumber
yourself with too long a line, as most do. And before
you begin to angle, cast to have the wind on your

back ; and the sun, if it shines, to be before you ;
and to fish down the stream, and carry the point or
top of your rod downward, by which means the
shadow of yourself, and rod too, will be the least
offensive to the fish, for the sight of any shadow
amazes the fish, and spoils your sport, of which you
must take a great care.'

Again, in the well-known 'Treatyse of Fysshynge
wyth an angle' written some time about 1450, and
attributed to Dame Juliana Berners or Barnes, the
reasons are given why trout are not *always* caught ; and
indeed the information contained is so comprehensive
as practically to cover all the misfortunes and worries
which beset an angler when in pursuit of his favourite
sport.

The 'reasons' are thus set forth under the heading of

The xij Impedymentes,

Whych cause men to take no fyche withoute other
commyn causes wyche may casuelly hap.   The first
yf your harnes [1] be not good and well made.   The
ij^d is yf ye angle not yn bytying tyme.   The iij^d
yf the fyche be a frayde with ye syzt of any man.
The iiijth, yf the watur be wery thilke or redde as
bye of any floyd falle late.   The v^d yf the fyche styr

---

[1] Harnes = harness = tackle.

not for colde or feyr. The vi[th] is if the water be
very hote. The vii[th] yf it reyne. The viij[th] yf hyt
hayl or snowe. The ix yf ther be any tempest of any
wether. The x yf hyt be a greyt wynde by any coste.
The xij yf hyt be by the northe or north est or
sowthe est for commenly nether by wynter nor by
sumer yf the wynde haue any parte of thys costes
the fysche wyll not commynly byte ne styre. The
weste and the sowthe be ryght good, yet of the two
the sowth is the bettur.'

Dame Juliana would seem to have forgotten the
eleventh ' Impedymente.'

Now these recommendations of Mr. Barker,
and the ' Impedymentes ' of Dame Juliana, comprise
nearly the whole duty of man as regards fly fishing,
and also the causes which operate against successful
angling.

What better advice can be offered to a beginner
than to fish ' fine and far off,' which to all intents and
purposes is what Mr. Barker says ; or that the fisherman must keep himself and his rod out of sight
as far as possible, as is also recommended ? The
greatest expert on Test-side cannot from the depths
of his hardly acquired knowledge advance any more
useful or practical suggestions. Again, that your
shadow should not be cast on the water is advice

of the soundest character; and that an unnecessarily long line should be avoided is an equally valuable hint. The only point on which modern anglers would probably join issue with Mr. Barker is his dictum that *down*-stream fishing should be practised.

As to the 'Impedymentes,' they, in the main, touch nearly every circumstance which in these days militates against a good day's sport, or, indeed, any sport at all, just as they did when the 'Treatyse of Fysshynge wyth an angle' was written over 400 years ago. The habits of trout seem, indeed, to have altered so little in the above-mentioned period that the modern fisherman, when driven to extremities by lack of success, may console himself with the reflection that things were much the same in the days of Henry VI. as in the reign of our present gracious Sovereign.

But when one turns to consider the question of what are the best rods and tackle to be used for the various kinds of trout fishing, then the task of venturing to advise becomes complicated in the extreme. It is not many years since trout-rods were generally manufactured of ash, hickory, lancewood, or greenheart, and of those woods much use is still made; but American anglers found out that strips of cane, carefully selected, built up and fastened by certain

glues or cements, were as good or better than the
rods composed solely of wood. Since then the manu-
facture and sale in England of split-cane rods has
been very large. A steel centre, which is supposed to
give additional flexibility and strength to the rod, is
now sometimes inserted. However, on some occasions people return to their old loves, and in the case
of cane rods this I believe is so.

Messrs. Farlow, who have kindly furnished me
with much valuable information on the subject of
what the public now require as regards trout-rods and
tackle generally, state that of late many more greenheart rods are being made, and that cane rods are gradually going out of fashion. The butts of these green-
heart-built 'limber gads,' as the North-country song
calls them, are fined down, and cork grasps fitted on,
thereby considerably decreasing the weight of the
rod. In fact, Messrs. Farlow consider the day of the
cane-built article nearly over. Certainly the prices
charged for rods made of cane are ridiculously large
as compared with those placed on wooden ones.
In addition, the cane rod requires much more
looking after and attention than does one composed
solely of wood. It has to be sent to the maker
every winter to be overhauled ; and though a really
good cane rod is a most valuable servant, it is a

troublesome one, and, moreover, is a rare thing to meet with.

It will be obvious that to fish properly such a river as, say the Test or the Kennet, a somewhat longer rod is required than would be the case if the Lea or the Mimram were the stream. For the Test is far broader that the Lea ; and though by deepish wading a somewhat greater quantity of water may be covered, a nine-feet-six-inch rod could not under ordinary circumstances be considered as effective as one of ten feet six inches or eleven feet. Therefore, as in the case of the choice of wading equipment, the only counsel that can fairly be given to the fishing novice is to go to a good firm of rod-makers and explain his wants to them. He will then in all likelihood be properly turned out.

Nevertheless while on the subject of trout rods there are certain suggestions and observations which may with due diffidence be offered for the consideration of fishermen. Taking first of all, then, the requisites of a rod to be used for fly fishing on rivers of an average size, such as the Derbyshire Wye or the Hertfordshire Lea, let us see what the length had better be, and of what material it should be made. Here we are at once confronted by the difficulty that every fisherman has a fancy of his own. The

writer knows one first-class angler who always uses a
nine-feet-six-inches greenheart rod, and declines to
avail himself of one of greater length.   On the other
hand, many equally good exponents of the art scoff
at such a sized rod, and affirm, with many strengthen-
ing adjectives, that nothing under ten feet six inches
is any use, especially if big trout are the quarry to be
caught.

As a matter of fact, the rod with which the fisherman can fish most accurately and with the least
fatigue is the one he should use ; and this knowledge
can only come from experience.   It does not matter
a drop of scentless paraffin whether the rod be composed of greenheart, hickory, or split cane, so long as
it suits its wielder.   It is of no consequence whether
it has cost seven guineas, as a cane rod sometimes
does, or whether one pound has been its price, so long
as it has plenty of power, is not too 'whippy' a fatal
fault is not too stiff and 'barge-poley,' is well
balanced, and not too heavy.   A rod thus fashioned
is a good one, of whatever material and by whatever
maker it may be built.

All that has been said with regard to single-
handed trout rods   for one must place in that category all rods up to eleven feet or thereabouts —applies
equally to double-handed ones, and therefore there is

not much to be added with respect to them which has not been already touched on in the above remarks on single-handed rods. A longer weapon than an eleven-foot rod must almost of necessity be double-handed, and under certain conditions of broad water it may be extremely useful. But for 'all round' trout fishing with a fly, ten feet six inches is, the writer ventures to think, as serviceable a length for a rod as can well be suggested.

It is most important for the fisherman to select his casting lines carefully, and not to buy too large a number of them at one time ; for fine casts, keep them as best one may, are not often to be trusted a second year. If during the winter they are put away in a box, wrapped up in paper and thus kept as far as possible from the influence of light and air, they may perhaps last for a second season; but when every precaution has been taken, there is a sense of insecurity about them, and the writer would advise the purchase of a new batch at the beginning of each spring. This may sound somewhat extravagant ; but what is more irritating and more likely to upset a fisherman's nerves and temper for the day, than to lose the first fish of the year by having put on an old cast, and to discover, too late, that it was rotten ? The few extra shillings which should have been ex-

pended on the purchase of a dozen new casting lines,
which expenditure would probably have obviated a
sad catastrophe, would not have been considered as
worth mention in comparison with such an irre-
vocable disaster. So the writer's advice is, do not
keep old casts, but get fresh ones yearly.

Of course there are, as always, exceptions to this
rule. One admirable fisherman—in fact, about the
best the writer knows—seems always to make up his
casting lines from old pieces of gut. A new cast in its
entirety is hardly ever to be seen on his line ; but the
secret is that in all probability he selects the sound
lengths out of his old stock. There are always some
odd pieces which seem to keep their freshness and
not to decay as quickly as the rest, and from them he
manages to concoct crafty casting lines, wherewith he
usually contrives to kill a good many more trout than
most other people.

Then there is the question whether casts should
be tapered, or of a uniform thickness ? A slight
taper is certainly an assistance, but if a batch of good
sound gut of the size required is placed before the
sportsman, he will be unwise if he declines it because
it is not tapered. Old Isaac says something on this
point : 'First let your rod be light and very gentle.
I take the best to be of two pieces, and let not your

11

line exceed, especially for three or four links next to
the hook—I say not exceed—three or four hairs at
the most, though you may fish a little stronger above in
the upper part of your line; but if you can attain to
angle with one hair, you shall have more rises, and
catch more fish.' This is sound and excellent advice.

But very fine gut, where trout run large and a
sufficient quantity of weed is left in the water, is a
dangerous luxury. It is possible that an extra fish or
so may be risen and hooked by using such a cast,
but the angler will probably be broken more often
than if he had on a slightly stronger quality of gut.
So matters would thus about equalise themselves.
Naturally, the finer the casting line, especially if the
weather be bright and glary and the water low, the
less chance there is of scaring the fish. But a cast
which can, under ordinary circumstances, hold a fair-
sized trout is equally a necessity; so, as in most
things, moderation in the matter of the thickness or
thinness of casting lines is to be advocated. They
are to be procured quite fine enough not to 'put
down' your fish, and at the same time of sufficient
strength to hold an energetic and well-sized trout.

That they should be thoroughly soaked before being
brought into active use is, as has been before said,
absolutely essential; and it is wonderful what a strain

really good gut, even if very fine, will stand, when
this precaution has been taken. Gut usually glitters
unduly if a slight staining be not given to it. Soaking it in warmish water into which a little coffee has
been poured is a sufficient cure for this evil, and a
quarter of an hour will be long enough for the im-
mersion ; a quiet brownish colour should be the
result. Sir Herbert Maxwell says that an alder leaf
rubbed down the casting line takes off all the shine,
and in its place a dull hue, about the tint required,
is left.

Casting lines are often made eight or nine feet
long. In my opinion, this, as a rule, is an unnecessary length. Six or seven feet of gut is usually
sufficient. Also, by taking off the extra two or three
feet from the cast, a most valuable spare length is obtained, which can be utilised when a breakage has
occurred, or when the casting-line has become too
short owing to a constant change of flies, each such
shift diminishing the length of gut by an inch or so.
It is always as well to have half a dozen fine tippets
in one's fly-book in case a slight addition to the length
of the cast is required.

After use the casting line should be placed between
the flannel leaves of the fly-book, as then it dries gradually, and is available for further work. A piece of

H 2

india-rubber in the fisherman's pocket is a good thing, for by rubbing it down the casting line all 'kinks' and bends are removed, the gut hangs straight, and becomes more easy to manipulate. To recapitulate. First, it is unwise to rely too much, if at all, on old casting lines ; secondly, it is unnecessary to have them of too great a length, as this is not essential, and adds to the difficulty of accurate casting ; thirdly, always soak gut thoroughly before using it ; fourthly, take care to test every cast before beginning to fish, and see that it has neither kink, bend nor unintentional knot in it. If these precautions are observed and the fisherman has by experience learnt what thickness of gut he requires, nothing more need be suggested on the subject.

Turning to the question of reels, the main points to be considered when purchasing one of these articles, are :—

1st. Its suitability for the rod on which it is to be used.

2nd. Its line-holding capacity.

3rd. Its durability.

As regards the first of these, a reel should be chosen of a weight which properly adjusts the balance of the rod. As to how much line it should hold, one large enough to contain sufficient for the general use

to which the rod will be put should be selected ; and
as to its material, it is unwise to buy a reel which is
composed of any brittle substance, such as ebonite. A
reel, to be satisfactory, should be one which works
smoothly, does not easily get out of order, and will
stand an occasional fall or hard knock.   I have therefore come to the conclusion that an ordinary well-
made bronzed winch of the 2½-inch size, holding some
thirty-five yards of line, is, take it all round, the best
for a 10-feet-6-inch or an 11-feet rod.   Ebonite is a
risky substance, liable to break if it has a fall ; brass
is cheaper, and answers well for ordinary purposes ;
wood, with a metal rim and metal inner revolving
plate, is occasionally used.

Touching the sort of line which should be selected
for trout fishing, nothing better can be suggested
than a waterproof, plaited, dressed silk one, tapered.
This seems as good as anything can possibly be,
and it appears unnecessary to say much more on the
subject ; the only additional suggestions being that,
when buying a trout-line, care should be taken to
choose one heavy enough to cast against the wind,
and not too weighty for the rod to throw com-
fortably.   Too heavy a line strains the rod and tires
and irritates the caster.   The length above mentioned,
viz. 35 yards, is ample for any ordinary purpose.   If

a double-handed 12 or 14 feet rod is being used, a larger reel and longer line are of course needed.

Having thus far dealt with the equipment of a trout fisherman, the only important matter which remains to be considered—and very careful thought is here required—is that of the flies with which he proposes to kill his fish.

From the days of Dame Juliana Berners and Isaac Walton down to the present time when Mr. Halford's beautiful work on artificial flies is the standard book on the subject, every angler has some dozen kinds of flies which seem to be universally acknowledged as indispensable, together with certain fancy ones of his own in which he greatly believes, but which could not, by any stretch of the imagination—even that of a fisherman—be taken as imitations of any natural insect. Take, for instance, the 'Toppy,' or 'Hofland's Fancy,' or 'Greenwell's Glory' (an admirable fly on northern streams), or the 'Zulu;' they are not supposed to be exactly like any insect that flies or crawls, but at the same time they possess a sort of resemblance to some, and to this semi-likeness the wile of man has added certain attractive colours to further beguile the trout.

Isaac Walton has said 'there are twelve kinds of artificially made flies to angle with on the top of the

water,' and having described them he winds up by
affirming that ' thus you have a jury of flies likely to
betray and condemn all the trouts in the river.'
Without going into the tying of the various flies he
recommends, it would seem that even now it is unnecessary to have more varieties in your box or book
than the number here mentioned.

The names of those flies which would seem to be
most generally useful—at least in the estimation of
the writer—are :—

The March brown.	Wickham's Fancy.
The Red-spinner.	The Black-gnat.
The Alder.	The Yellow-dun.
The Olive-dun.	The Hare's-ear-dun
The Red-quill-gnat.	(with gold ribbed body for
The Olive-quill-gnat.	choice).
The Governor.	The Iron-blue dun.

This of course is only a list of 'general utility'
flies which I would suggest should always be in the
possession of the angler.

Every river has its own private list, which should
usually be carefully studied. For example, in Derbyshire, where would one be without having in one's
fly-box some ' Bumbles,' and Ash-duns, and Apple-
green-duns ? Or on the Test when the ' Grannom '
is ' up,' without the exact, or as nearly exact as pos-

sible, representation of that fly, to put on when it makes its annual appearance? It is needless to add that during the brief but ecstatic period when the Mayfly arrives the fisherman must have many deftly constructed specimens of the fly-tier's art at his disposal, should he be fortunate enough to find himself fishing waters where that mysterious insect exists.

It would be out of place in this book to do more than touch very slightly on the question of flies, artificial or natural. The subject is one which has been so fully and ably dealt with by fishing experts for many years past, that any extended disquisition must be superfluous. But the list which has been given above will, as a general rule, be found sufficient to ensure success on most English or Scotch streams should the trout be at all inclined to rise.

It is rare, nowadays, to find a dry-fly fisherman using any other flies than those tied on ' eyed ' hooks.

The reasons for this are obvious. In the first place, ' eyed ' hooks are much less liable to ' crack ' off when casting than hooks tied on gut. In the second place, as there is no gut to spoil and get rotten by keeping, the flies last very much longer, in fact till they are either worn out by use, or lost in a fish. These are material advantages which at once explain the popularity of eyed hooks.

Whether the hooks have the eyes turned up, or down, or straight, does not seem to matter much, and is to a large extent a subject of personal predilection.

The process of tying an eyed fly to the casting line is both simple and rapid. There are several ways of doing this, but the most effective as well as the simplest seems to me to be the following :—After the casting line has been attached to the reel line, slip its end through the eye of the hook, then form a loop of the end of the casting line, and tie a single knot round the cast itself.

If this is carefully done, not more than half an inch of the casting line need be cut off and wasted. The next thing is to push the slip-loop thus made back *over* the fly, taking care the wings are not damaged in the process ; then gently pull and jam the knot tight into the eye. With a little practice, and with properly soaked gut, this can be done with hardly any trouble, and much speed.

The various sizes of hooks are shown in fig. 7, with their descriptive numbers, on the scale usually accepted.

There are many small additions to an angler's equipment which may be mentioned, not perhaps as 'necessaries,' but as being useful on occasions. For instance, in these days most dry-fly fishermen

take with them a small bottle of scentless paraffin
oil, with which, by the aid of a tiny brush, they just
touch the body, hackles, and tail of the fly they
intend to use. This adds very largely to the floating
capabilities of the artificial insect, and lessens
enormously the trouble and fatigue of drying the fly,
after having put it over a fish, by means of a dozen
throws in the air before again allowing it to float over
a possible victim.

Others, again, place a bit of dried deer's fat in

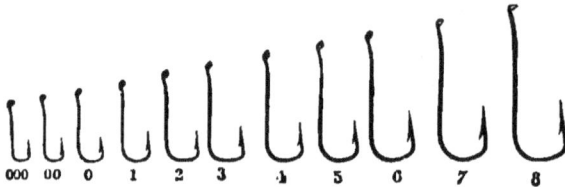

000   00   0   1   2   3   4   5   6   7   8

FIG. 7

their bag, wherewith to rub their reel line to prevent
it from 'kinking;' while some keep their spare casting
lines and gut lengths in a piece of soft leather which has
been anointed with vaseline. Some like one form of
cutter to take off the gut ends when the fly is tied on,
and some another. In fact, in fishing as in most
other pursuits, every one who is really keen will discover for himself endless little fancies and hobbies
which amuse and interest him, and do no harm to
any one else.

There are, however, certain rules with respect to all trout fishing, whether with fly or worm, which should be observed by every fisherman. The first is, when fishing for trout to keep as far from the bank and as much out of sight as possible; the second is, to persevere, remembering the old saying that 'He who keeps his fly on the water longest kills most fish.' The third is, never to be in a hurry; more fish have been lost from this cause than from almost any other. If he who goes a-fishing will remember these general principles, and each time that he is by the riverside will carefully study the peculiarities of the fish on that particular day, and the variety and appearance of the natural insects which hover around him, he will soon become a master even among those who have always regarded with enthusiasm, and practised with success, the most enthralling, graceful, and delightful art of trout fishing.

The question what is the best weather for fishing is one which has been warmly discussed by anglers ever since rods and lines were invented; but we seem to be as far off as ever we were from the acquirement of accurate knowledge on this subject.

Perhaps it is as well that this should be so, for what a depressing affair fishing would become if, by a glance at the weathercock, and another at the thermometer before starting, we could make certain of the amount of sport we might expect ! Mercifully, however, this is not the case; for very often a day which was seemingly as unpromising as could possibly be, has in the end proved one of the best of the season.

The 'Treatyse on Fysshynge' to which allusion has before been made, says: 'Ye schall angle as y seyde befor in darke lowryng wedur when the wynde blowethe softely and yn somer seasen when hyt ys brennyng hote.' Thus Dame Juliana. Isaac Walton remarks on this subject that 'You are to take notice that of the winds the south wind is said to be the best. One observes, that

> ——when the wind is south
> It blows your bait into a fish's mouth.

Next to that, the west wind is believed to be the best; and having told you that the east wind is the worst, I need not tell you which wind is the best in the third degree: and yet (as Solomon observes) that "he that considers the winds shall never sow;" so he that busies his head too much about them, if the weather be not made extreme cold by an east wind, shall be a little superstitious ; for, as it is observed by some that "there is no good horse of a bad colour;" so I have observed, that if it be a cloudy day, and not extreme cold, let the wind sit in what quarter it will and do its worst, I heed it not.'

These are the views of two of the ancient writers on angling, and old Isaac's quaint observations are full of sound common sense.

The older one grows, and the more one fishes, the greater appears the folly of anticipating bad sport by taking to heart in too large a degree the atmospheric conditions which surround one. Naturally, from a physical standpoint, it is far pleasanter to go a-fishing with a warm, soft, south-west breeze blowing gently up-stream, than to have to force a fly against half a gale of wind whistling from out the bleak north-east. But though in the one case you may be kept at a comfortable temperature all day, and in the other be bitterly cold, it is by no means so certain that the north-easter will not bring a better bag with it than the south-wester.

On this point it is worth while to refer to the fishing diary ; and it is interesting to note what good days have been obtained, as to which the entries concerning the weather run somewhat on these lines : 'Strong, cold, down-stream wind from N.E. : Hailstorms at times : Fish rose well between the squalls.' Not only does this apply to the earlier spring months, when inclement weather is to be expected, but also to May and June, when cold 'snaps' occur at the most unexpected moments.

Sometimes, when fishing during the Mayfly season, I have noticed that provided the strength of the wind be not too great, the trout seem to rise at the fly better when the breeze is *down* stream than when it blows in an *up*-stream direction.

Why this should be, I am not quite clear ; unless it is that the artificial Mayfly floats more naturally when being driven down-stream by the wind, instead of being somewhat checked in its movement by a contrary breeze. If this is so, then, as far as I can judge, a *light* breeze blowing the same way as the flow of the river may often be preferable to one which would as a general rule be considered more favourable ; this, at any rate, when the Mayfly is about.

Of course *very* cold weather is inimical to much sport : so is very sultry weather, and in this respect extremes meet. But in the case of the former condition, should the sun manage to penetrate the clouds and touch the water with his gleams, very likely some dark duns will hatch out and the trout take them eagerly. However, as I have said before, it is little use speculating on the weather, and so fish on whatever it may chance to be.

There is only one of Nature's productions which

always seems to be fatal to fishing (naturally I except a 'spate,' and 'coloured' water), and that is a cold, dank mist, which rises and envelopes the river like a shroud. This seems to put an end to all sport, and when it appears the angler may pack up and go home.

The question may be asked, 'In what months and at what times of the day during those months should fishing be undertaken?' Well, in Devonshire and certain other parts of the country—but in few, I am glad to say—I believe trout fishing begins in some rivers on February 1, and on others on March 1. It is possible that in some districts trout *may* be in condition as early as this, but I do not think that trout fishing ought to begin before April 1 at the earliest. By then the small fish in Devonshire waters may be fit for catching, but it is rare to find any trout in the larger South country rivers which are in condition much before May. As a general rule, October 1 is the date for trout fishing to cease, and this would seem about the proper time ; at any rate it should not be later. During the earlier weeks of September I have often caught trout in lovely condition, but soon after this date the fish begin to 'go back,' and then it is time to put away one's rods and lines, at any rate as far as trout fishing is concerned.

CHAPTER VI

SOME HEAVY BASKETS AND BIG FISH

IT will be easily understood that it is not possible to do more than give a very imperfect account of remarkable baskets of trout made during the last thirty years. In that period a vast number of good days' sport have occurred which could fairly claim their right to be enrolled amongst fishing records, or would at the least be deserving of detailed remark, of which the fortunate fisherman himself and his intimate friends are alone cognisant.

It would, therefore, be obviously a hopeless task to attempt anything in this direction, beyond jotting down some few results which have come under my personal observation, and alluding to some baskets of trout, details of which have been supplied me through the courtesy of Mr. R. Anderson, of Princes Street, Edinburgh; Mr. W. Thomas, secretary of the Fishing Club at Mullingar; Mr. W. G. Craven, and other kind friends.

With regard to English rivers, two days' fishing on the Mimram, in Hertfordshire, are well worthy of notice. In the year when they took place, 1879, this little river held large trout, as in certain portions it does still. The two days' sport to which I allude occurred on July 2 and July 31 on the Hoo water belonging to Lord Dacre, Lord Romney being the fisherman. On July 2 he killed 8 trout, which together weighed 25 lbs., and turned back two more. On July 31 he caught 4 trout weighing respectively 5$\frac{1}{4}$, 5, 4$\frac{1}{2}$, and 3 lbs. ; total 17$\frac{3}{4}$ lbs., or an average of more than 4$\frac{1}{4}$ lbs. On this day Lord Romney returned four other fish. These trout were killed with a small 9-foot greenheart rod and a fly something between a coch-y-Bondu and a 'hackle' alder. A southerly wind was blowing, with heavy showers at intervals. Now, without asserting that this latter day is an absolute 'record,' it is one which is hard to beat, as far as fly fishing for ordinary *Salmo fario* in England is concerned, and for this reason it is here mentioned.

The number of trout of 4 lbs. and over which are caught during the course of the year in the Kennet, Test and Itchen is not very great ; but at the same time each season brings news of some monsters of 7 lbs., and even larger, having been captured.

I

Up to a short time ago the Hertfordshire Lea always principally during the Mayfly season—furnished a considerable supply of trout weighing between 2 and 5 lbs. I once saw placed on the floor of a railway carriage four trout averaging well over 3 lbs., the proceeds of a day's fishing in the Mayfly time on the Hatfield Club's water on that river; and the fortunate angler told me that other rods had done as well, or better. Reverting to the Mimram, of recent times no stream, taking its size into consideration, has afforded more sport, or yielded larger fish. The Tewin water length, near Welwyn, has always held very big trout, and still preserves that sound characteristic. No weightier bag is often heard of than that which was made there a few years back by two friends of mine, when they eventually took to returning to the river every trout they caught which did not seem to be well above 2½ lbs. !

The Midland and Northern streams do not produce trout of the same size as do the Southern rivers; and on the Derbyshire Wye a 2-lb. fish is considered a very good one. Nevertheless I know of one angler who in the season of 1897, on the stretch of the 'open' water of that river, between Rowsley and Bakewell, caught three trout each of which weighed over 3 lbs.

In the smaller Devonshire rivers a trout of 1 lb. is regarded as a very worthy fish indeed.

Turning to Scotland, Mr. R. Anderson has supplied me with some notes and observations on large fish killed north of the Tweed. He mentions first of all the Big trout—it deserves a capital 'B'—caught in Loch Stennis in Orkney in March 1889. And here I should observe that in writing about Scotch trout records, the captures were effected not solely by the fly, but by spinning, by set lines, and in fact by every sort of piscatorial device. This Stennis trout weighed 29 lbs., measured 38 inches in length, and was 24 inches in girth. It was taken on a set line, but the nature of the bait is not stated. There has been a doubt expressed whether this fish was a true trout, but that is a point I am not able to solve. I tell the tale as it was told to me.[1] The largest river trout which has ever come to Mr. Anderson's personal knowledge is one which was killed in the Tay by Colonel Dobiggin on September 10, 1842. That gentleman was fishing with fly for sea trout on the Murthly water, and in the Birnam Burn pool he hooked the fish. After a long fight, John Miller, the fisherman,

[1] Since the above was written—in fact on June 28, 1899 I have heard from Major Levett, of Rowsley, respecting this fish. He says: 'I was intimately acquainted with the big Stennis trout you wrote about in *Fur, Feather, and Fin.* The fish was caught on a trimmer set with a live bait on a cod-hook tied on to loose horsehair.' G.

landed it for him. It was a male fish weighing 27½ lbs. Mr. Anderson's father—with whom as well as with John Miller I was well acquainted—was present on the occasion, and took an outline of it on paper. Its length was 39½ inches, and a rough sketch of it is now in Mr. Anderson's possession.

A trout of 27 lbs. was killed by Mr. H. Almond in the lower part of the Inver, in October 1869. It took a small 'Childers' fly, and was 41 inches in length.

Mr. Anderson informs me that he is cognisant of the following big trout having being killed on Loch Rannoch, Perthshire : one of 18½ lbs., caught by Mr. J. Watson Lyall, trolling with the natural minnow; another of 17½ lbs., killed by the late Mr. David Cox, of Lochie. This was taken on a small lake fly, with a 12-foot rod and a short line. In consequence, a prolonged struggle took place, and the fish was only ultimately secured by very careful handling of the boat. Mr. James Elliott caught one of 17 lbs. This fish took a small phantom minnow which was trailing behind the boat, while the men were pulling hurriedly into a bay to escape the violence of a storm which had suddenly sprung up. It was during the first week of September that this occurred.

A very fine trout was taken in Loch Rannoch in the summer of 1897, Mr. J. H. Fergusson of Surbiton being the fortunate captor. The fish weighed 13 lbs.

It was 28 inches in length, and 21 inches in girth. It may be added that Mr. Fergusson is in his 84th year !

On Loch Ericht, in Perthshire, the late Struan Robertson killed in July 1855, during one day's trolling, three trout which weighed together 53 lbs. One of them weighing 17 lbs. is set up, and is now in Mr. Anderson's possession.

The late Sir Molyneaux Nepean, in July 1875, killed 83 lbs. weight of trout on this same loch. The five largest together weighed 33 lbs. They were all taken with either phantom or natural minnow.

A lady, Mrs. Archibald, had a wonderful day's sport in May 1893 on Loch Merkland, Sutherlandshire. Mr. and Mrs. Archibald were fishing in separate boats. Soon after they started the breeze died away, and after waiting for some time Mrs. Archibald decided to try trolling, which she did by attaching a large phantom minnow to the line on her 10-foot fly rod. As nothing happened, she soon reeled in, fastened a small trout which she had previously caught to one of the triangles of the phantom, and again started trolling. Almost at once the bait was seized by a big fish. Mr. Archibald pulled to his wife's assistance, and eventually she landed a grand trout of 15 lbs. weight. Nor was this all : for soon after she had recommenced fishing, with the same

kind of bait and tackle she killed another trout of
8½ lbs.

Mr. Anderson writes further that ' Beside the
fish I have mentioned, I have known of numbers of
others taken in these lochs of 20 lbs., 17, 15, 14,
13, 10, and 8 lbs. Perhaps other large lakes would
yield as good results if they were as systemetically
fished. The late Sir Everett Millais informed me
that on Loch Ness he had killed, amongst other large
trout, one of 14 lbs. Another gentleman in a month's
trolling there, caught, in addition to other trout, thirty
that weighed from 5 to 10 lbs. each.

' My brother and I, in a small loch in Perthshire,
in June 1875, killed in one day, 93 trout that weighed
65 lbs. In the same loch a friend of mine and I
killed 114 trout of 73 lbs. total weight.

' On the Murthly water of the Tay, I killed one
day at the beginning of June 1868, fishing with the
fly between the hours of 11 A.M. and 5 P.M., over
50 lbs. weight of trout. I took home and weighed 53
trout, which together came to 46 lbs., and gave away
six which must have scaled fully 4 lbs. The two largest
fish were respectively 3 and 3½ lbs. On another occasion, in May 1871, when fishing the Kinnoul
water in the Tay, my brother and I killed 43 trout
weighing 45½ lbs.

'Mr. John White, manager of the Loch Leven fishings, has given me a list of the best baskets made on Loch Leven during the last ten years.

' Underneath I give some of the best of them, only the "takes" given do not represent *individual*, but *boat* catches.[1]

' In 1887 there were 12 baskets ranging from 35 to 50 lbs. ; the trout averaging over 1 lb. each. The best were :—

1887. April 21, 48 trout weighing together	56 lbs.
„ „ 29, 41 „ „ „	50 lbs.
„ July 28, 46 „ „ „	47 lbs.
1888. June 2, 53 „ „ „	49 lbs. 8 oz.
„ „ 7, 60 „ „ „	56 lbs. 8 oz.
„ „ 7, 70 „ „ „	53 lbs. 12 oz.
„ July 13, 56 „ „ „	57 lbs.

'On June 2, fishing in the Kinross competition, Mr. Malloch of Perth killed 52 trout of 44 lbs. 6 oz. weight.

1889. Aug. 12, 30 trout weighing together	44 lbs. 6 oz.
„ „ 16, 45 „ „ „	38 lbs. 6 oz.
1890. May 16, 65 „ „ „	47 lbs. 8 oz.
„ „ 50 „ „ „	34 lbs.
1891. May 24, 55 „ „ „	40 lbs.
„ „ 30, 44 „ „ „	36 lbs. 8 oz.
1892. May 25, 54 „ „ „	43 lbs. 8 oz.
„ „ 26, 62 „ „ „	52 lbs.

[1] I conclude this means two persons fishing in one boat.

1892.	May 31, 44	trout weighing together			40 lbs. 8 oz.
1893.	May 2, 47	„	„	„	39 lbs.
„	Aug. 22, 69	„	„	„	49 lbs. 12 oz.
„	„ 50	„	„	„	37 lbs.
1894.	June 4, 55	„	„	„	39 lbs.
1895.	Aug. 24, 26	„	„	„	36 lbs.
1896.	May 21, 33	„	„	„	29 lbs. 12 oz.
„	July 30, 29	„	„	„ .	31 lbs.
„	Aug. 17, 24	„	„	„	26 lbs.
1897.	May 16, 36	„	„	„	39 lbs.'

With regard to the large trout whose various captures in different lakes are above recorded, Mr. Anderson says : ' I consider Mr. Almond's big trout which was killed in the Inver to be a pure *Salmo fario*. The large trout taken on the Murthly water of the Tay was also a *Salmo fario*. My father, who was present, took a full description of it, and sent it to the leading naturalists of the day, and they all agreed it was just a fine specimen of the ordinary river trout. The other large trout I mention were sent to us as examples of the *Salmo ferox*.'

I take the following from the ' Field ' of October 23, 1897 : ' What is believed to be the biggest Tweed or Teviot trout on record was taken on Saturday afternoon last.

' Mr. W. Shiell, of Kelso, fishing with bait, had a basket weighing about two stone : and amongst other fish was a large yellow trout caught in the Teviot above

Sunlaws Mill Cauld. This was a male fish not by any means in the best of condition.[1] It weighed 7¼ lbs., and measured 27½ inches in length; 13½ inches in greatest girth. Some doubt was expressed when it was exhibited in Mr. Steel's window whether it was not a bull trout; but judging from the recognised peculiarities of structure and the profusion of black and red spots with which it was covered all over, the balance of opinion by competent authorities is that it was a genuine freshwater fish. This being so, it is the largest trout ever taken from any of the rivers in the Kelso district.'

I have mentioned that in even the best English rivers the number of trout of over 4 lbs. in weight, which are caught in any one season, is not very great; and with regard to this I have received from Mr. W. G. Craven much interesting information respecting the well-known Longstock Club water on the Test. In addition to the particulars as to the weights and numbers of the trout caught, some very curious details are given as to the formation of the Club in question, and the methods of fishing practised a hundred years ago in that part of the Test.

[1] Was it likely to be at that time of year? But those Tweed fishermen are past praying for!

Mr. Craven says: 'In all probability the Test has more club and private waters than any other stream of its size in England; and moreover has older records—which records have been, and are, properly kept by such clubs as the Longstock and Stockbridge ones—than any other river.

'The Longstock Club is the same as the old Leckford Club, though the members have crossed the river for the sake of a more convenient house, and have thereby lost a small piece of the Leckford water.

'The Club records date from 1798, and from then till now every fish caught has been duly weighed, and the books kept in perfect order.

'It is a fact that, for more reasons than one, the "catch" deteriorated from 1850 to 1870, compared with that at the commencement of the century; when over 400 fish were caught during one Mayfly season in the water below Testcombe Bridge to within three-quarters of a mile of Stockbridge.

'The Longstock Club water runs entirely on a chalk bed, and there is no gravel except under the hills on each side of the valley. The bottom of the river is, in consequence, all mud collected from the peaty soil through which it runs. The course of the main stream has also completely changed in the centre of the valley, in consequence of the artificial irrigation of the meadows. This, has, however, considerably increased the amount of fishing water, as all the carriers, or "carriages," as the local folk call them, hold trout.

'Whether to the clearness of the water, or the higher education of the fish, can be attributed the diminution of the size of the bags, it is impossible to say. But it is probable that with the increased number of sportsmen almost every trout has been either "pricked," or hooked and returned to the river as not "sizeable."

' Possibly the "patchiness" of the Mayfly may have had something to do with this, but at any rate we are now told that 100 to 150 fish during a whole season, Mayfly time included, may be considered a satisfactory number: 9 to 13 trout in a day to one rod being sometimes the red-letter bag of the season. These fish would average 2 lbs. in weight. As to big fish of 5 lbs. and upwards—5 lbs. 8 oz. being the largest killed – they amount to a dozen in 100 years. Four-pound fish have been more numerous, possibly amounting to a hundred in the same period. The 10,000 or 11,000 trout caught, average within a fraction of 2 lbs., so that it is clear fish of 2 and 3 lbs. must be the most numerous.

'Till about the year 1868 the only flies used were
large ones, tied on No. 8 hooks,[1] Mayflies and moths
being the most common artificial ones.   Early in the
century the "blow line" was introduced from Ireland, the natural fly on a No. 8 hook being used.
For this purpose a hollow cane rod of some 24 feet,[2]
or even one made of deal, was the general weapon.
Nowadays such a proceeding is hardly ever heard
of in English rivers.   About the year 1860 the tiny
"dry" artificial small gnat, and the 10 or 11 foot
rod, which had previously come into use on other
smaller streams in England, became the customary
system of fishing in the Test.

'When this river has passed Stockbridge, its bed
again becomes of a gravelly nature, but not in so
great a degree as above Wherwell, or the upper reaches
belonging to Lord Portsmouth and Mr. Portal.   Below
Romsey trout run to a larger size.

'The Stockbridge Club I believe to have been
founded by Hampshire men.   Not so the Leckford,
now the Longstock Club.   This was created by gentle-
men from Northumberland, who came down solely
for the Mayfly season ; the Rev. — Ogle being the
absolute originator, and first Secretary.

[1] See fig. 7, p. 106.
[2] Is this not an excessive length ?

'At that time Ogles, Sheridans, and the Duke of Argyll figure among the members of the Club. A little later come more Sheridans: while Abercorns, Streatfeilds, Bathursts and many others have up to these times filled the roll of members.'

These notes of Mr. Craven's afford much ground for consideration and reflection ; and the information he has supplied will, I am sure, be appreciated by the fishing community. Personally, I am not cognisant of any other fishing club which has existed without a break for 100 years, and has preserved, unbroken, its records and its history ; though of course there may be others which have done so. Northern anglers on Coquet side and in Yorkshire have, I am well aware, old-established fishing societies which may be able to furnish as ancient and well-authenticated reminiscences. But if this is so, it in no way affects the interest of Mr. Craven's account.

In truth it must have been no light or easy matter, at the end of the last century, to travel the whole way from Northumberland to Hampshire ; and men must have been keen sportsmen indeed to undertake such a journey for the sake of a week or ten days' fishing, even though it were the Mayfly time.

No doubt sport was excellent, as has been shown, and they were well rewarded for their energetic proceedings; but nevertheless it must have been an arduous undertaking.

What a difference there is nowadays! The 9.15 A.M. train from Waterloo will take you to Stockbridge, or anywhere in those parts, in time to allow you to begin fishing by 12.30, or thereabouts, and to return to London the same evening.

Undoubtedly in 1800 there did not exist the same conditions of hurry and 'rush' which now seem to pervade the country. It was probably as convenient to put on one side a week in which to make a journey, as it is now to arrange to run down to Derbyshire, fish for two days, and return to London by the night train, after fishing on the second day.

Instead of only having time to fish for two days, the angler of 80 or 90 years ago would have been easily able to spare a week or a fortnight in which to enjoy his favourite sport. At that period the affairs of life were taken more quietly, and very likely were conducted to their termination more thoroughly and effectively than is now the case. These Long-stock records go far to prove that in English waters big trout of 5 lbs. and upwards are somewhat rare fish.

Occasionally one sees a paragraph in a newspaper telling of the capture of some monster of 10 or 12 lbs., or of even a greater weight; but these are nearly

invariably, so to speak, 'hand-fed' fish. They are
trout which have for years inhabited some deep mill-
head or heavy water close to a house, where they
have been regularly fed on scraps and leavings from
the kitchen ; or have dwelt near the mouth of some
drain where they fattened on sewage relics. And in
general it is not by the agency of any artificial fly
that these great fish have died. By worm ; by pieces
of bread or other bait have they been captured ; and
thus they hardly come into the category of English
trout which have been fairly killed.

The late Mr. Frank Buckland, in his 'Natural
History of British Fishes,' recounts how in February
1868 some one sent him from Alresford a huge trout
which weighed 14 lbs. The model of this fish was despatched to the Fish Museum in South Kensington.

It was taken—*how* is not stated—in the act of
spawning in a small stream running through the
shrubberies at Alresford. Mr. Buckland carefully
weighed the fish, which was 30½ inches in length with
a girth of 19 inches at the middle of the abdomen.

He likewise mentions a trout which Lord Dorchester wrote about. This was also 14 lbs. weight,
and was taken near Winchfield.

Mr. Buckland further states in his notes on
Thames trout, that a friend of his, Mr. Forbes of

Chertsey, when spinning in Chertsey Weir, caught a fish of 16 lbs.

There was captured—also by spinning—in the Thames near Reading on April 19, 1880, a grand trout of 16 lbs. 15 oz. After this fish had been cast by Mr. Buckland it was sent to Windsor as a present to Her Majesty.

But to return to smaller trout caught by the dry fly.

The following notes from his fishing diary have been supplied me by a well-known angler who usually fishes the Itchen.

It would seem that in 1887, on May 20, his basket consisted of 13 trout weighing 25 lbs. 8 oz. The largest fish was 3 lbs Several were lost, and the rise lasted from one o'clock to four. The fly which killed was a small Olive-quill. The weather was cold, windy and showery.

On May 21 of the same year his bag was 10 trout, which weighed 20 lbs. There were two fish of 2¾ lbs. each. A small Red-quill took them. The rise occurred between two and half-past five. The weather was wild. A heavy thunderstorm took place about 2.30 P.M., accompanied by much hail and rain.

With regard to these two days, it is worthy of remark that such good sport should be obtained in the midst of weather which was cold and thoroughly unseasonable. This goes far to prove that no one should bother his head when he goes fishing as to what 'airt' the wind may be in, or what the weather may be, so long as he can fish the river properly, and the water itself is in 'order.'

But to return to the Itchen. This same gentleman, fishing on July 14, 1892, killed 12 trout weighing 23 lbs. The best fish weighed respectively 3½ and 3 lbs. 1 oz. The wind was N.E., the rise being between eleven and half-past three. Four fish were killed between eight and a quarter to nine in the evening. The flies used were a small Red-quill, and a Pink Wickham's Fancy. (Here the 'Fancy' fly comes in.) My correspondent adds : 'These are the three best days I have ever had with the dry fly ; but I always call 10 lbs. weight of trout a good day, whether with a " dry " fly and large trout, or a " wet " fly and small trout in the North, at any rate in any water which, though well preserved, is often fished.

' I well remember one fish of over 8 lbs. being caught with a fly in the Itchen one evening. This trout, however, lived just below a town and a mill, and was not therefore a fair test of what a fish might grow to in the open river.'

These Itchen reminiscences, coming as they do

K

from the pen of a first-rate and sedulous fisherman, give a good indication of what may be expected when dry-fly fishing on a South-country river, at least at any time of year except that when the Mayfly reigns supreme.

With regard to the attractiveness of that insect, the following information has been given me by Mrs. Ogden Smith, who is such a skilful creator of the beautiful artificial flies which so many of us use. She writes : ' In the early Fifties my father fishing in the Kennet at Hungerford, landed in one morning's fishing no less than 7 fish, the smallest weighing 2½ lbs., the largest 7 lbs. This was with a large dubbing-bodied May-fly.'

Ye gods ! that must have been a sight, indeed !

I myself, when fishing the Savernake length of this river, have often seen veritable monsters therein, but have never caught anything there over two pounds, though once I hooked an enormous fish I saw ' tailing ' on a shallow below a mill. He towed me about for some time, but finally the alder came away from his mouth, and I was left lamenting.

From what I have seen of the Kennet I am confident that on a really killing day it should still be practicable to secure a very heavy basketful.

Turning to the Northern Midlands, the following

day's sport which has there come under my own observation may perhaps be considered worthy of a
brief notice. It took place on June 8, 1897. Mr.
T. D. Croft and I killed between us sixty-nine trout, of
which number he claimed forty-nine. These forty-
nine were caught with the Mayfly in a little Derbyshire
stream called the Bradford ; the twenty I killed were
taken with the ash-dun and blue-dun on the Lathkill, into which river the Bradford flows. The fish
were a good Derbyshire sample, averaging well over
half a pound each. I see from my diary that there
was cold rain and wind on this day.

It is a curious fact that for the last two seasons
the Mayfly has become almost a thing of the past on
the Lathkill, though on the Bradford it exists in as
great quantities as formerly. This explains why the
trout I caught on this occasion were nearly all taken
on duns, while Mr. Croft's succumbed to the fascination of the Mayfly.

It is the more remarkable, as the two rivers in
question join at a point where the fishing is usually
considered to be as good as anywhere in the district ;
and yet of recent years as soon as ever the Bradford
becomes absorbed by the Lathkill, the Mayfly is
absent ; though there is, as far as can be discovered,
no difference whatever in vegetation, weed, river-bed,

water, or other environment necessary for the May-fly's existence.   Why this should be so I cannot for the life of me imagine.   But that it *is* the case I most regretfully have to confess, and so I must leave the solution of this mystery to wiser heads than mine.

CHAPTER VII

GRAYLING FISHING

GRAYLING (SALMO THYMALLUS)

THOUGH this fish is not generally held in such high estimation as is the trout, it possesses many merits of its own. For one thing, grayling come into condition about the end of July, and are to be caught until Christmas, thus taking the place of trout during the autumn months. Just when trout begin to 'go back' in condition, grayling reach their prime. In October, trout are valueless for eating purposes, whereas grayling are then excellent; and in November even better. Moreover the grayling, with his bright silvery appearance and seemly shape, is a very attractive fish to look upon.

From a purely piscatorial point of view indeed, the grayling is a very worthy object; for though it may not fight quite as hard as the trout, it is to be fished for in very similar fashion, and in rivers of

the same character.   It is found in Derbyshire,
in the Dove, Derwent, and Wye ; in Hampshire,
in the Avon, Test, Itchen, and other rivers; in
the Teme, Lugg, Arrow, and other Western streams.
There is magnificent grayling fishing in the rivers
which run through Wilton, and many are caught in
certain Yorkshire waters.   In Scotland, too, the
Clyde, Tweed, and Teviot now hold many of this
fish.

Grayling are very rapid risers, and have always
seemed to the writer to be more difficult to hook
than trout.   The number of grayling pricked and
missed in an ordinary day's fishing is usually very
large.   Their mouths are extremely delicate, and in
consequence light handling when playing them is
most essential.   With respect to the best flies for
grayling fishing, the ordinary autumn duns are always
useful.   In Derbyshire the ' bumble ' is a very killing
fly, tied with either an orange or a crimson body ; the
ash-dun and apple-green-dun are also largely used
in that county.   A tiny red tag, or a bit of bright
tinsel, added to an ordinary trout fly, is sometimes a
great attraction to grayling.

At Leintwardine, the science of ' grasshoppering '
has, I understand, been brought to great perfection ;
but, as I have never seen grayling fished for except

with the artificial fly, I cannot attempt to describe any other method.

When in a rising mood, grayling are apt to lie somewhat deeper in the water than do trout, and are also as a rule more inclined than trout to take the fly when it is just submerged.

They move about in company rather than by themselves, and according to the late David Foster of Ashbourne, who knew more of the habits of grayling and the best ways of catching them than almost anybody else, they are 'to a certain extent gregarious, generally frequenting the stills even when surface feeding. It is owing to this fact that they are given to rise nearer the rodster than the trout, ascending often from the deepest part of the river to seize a passing insect.'

When fishing for them, every part of the river should be tried; for what Mr. Foster has said is undoubtedly correct—namely, that grayling are more likely to rise in the deep, still reaches than trout would be, especially should the water run quietly close to the banks, when such places should always be fished carefully and enquiringly.

The grayling is a bold rising fish, though wonderfully apt to come 'short,' and for this reason is to be esteemed ; for whereas a trout does not, as a rule, rise fairly at the fly more than two or three times, and then seems to become disgusted with both you and it, thereafter declining to have anything more to do with you for some time, a grayling will, on the other hand, continue to rise for nearly as long as you like to persevere, if you now and again give him a minute or so between the casts.

Isaac Walton knew of this peculiarity, and says of the grayling that 'he bites not often at the minnow and is very gamesome at the fly and much simpler, and therefore, bolder than the trout, for he will rise twenty times at a fly if you miss him and yet rise again.'

The grayling when first taken out of the water has a peculiar smell, which in a limited degree resembles that of wild thyme : hence its name *Salmo thymallus.*

Referring for one moment to the subject of 'grasshoppering' alluded to above, it should be pointed out that the so-called grasshopper is an artificial bait not in the least resembling the natural insect, but rather some kind of fat, wingless grub. Grayling are also caught with gentles, or with small red worms, much in the manner adopted when worm fishing for trout.

With regard to size, in Derbyshire a pound grayling is a good fish ; in Wiltshire and Hampshire they run much larger, two and three pound fish being frequently killed.

As to what hours in the day are the best for catching grayling, it is obvious that, as they are in season from August to December, they rise at very different times during those months. In August I have seen them rise best, as a rule, from four o'clock to seven or eight; but as the days shorten they only feed when the chill of the morning is off the water, and cease when the evening damp and cold comes on. At those periods, therefore, the middle of the day is the most profitable time.

Take him altogether, the grayling is a fish well worthy of the angler's careful attention, and as he is to be caught in months when trout fishing is prohibited, he deserves perhaps higher commendation than he has hitherto received at the hands of the fly-fishing fraternity.

TROUT BREEDING

BY

COLONEL F. H. CUSTANCE

CHAPTER 1

HISTORICAL

PISCICULTURE has been described as 'the art of
fecundating and hatching fish-eggs and of nursing
young fish under protection till they are of an age to
take care of themselves.' Complete, however, as this
definition may have been considered thirty years ago,
it does not, as we shall presently see, embrace all that
is included in the scope of fish culture as understood
and practised at the present time. The origin of the
art seems to be lost in remote antiquity. It was in
China, we are told, that gunpowder was first used ;
and it is to the Celestial Empire also that we must
go for the earliest examples of fish culture.[1] Straw
and grass are, as we learn, tied round posts floated
against the current of a river. Hurdles, mats and
faggots, too, are attached to wooden posts fixed in the
bed of the watercourse. The fishes' spawn floating
down with the stream is arrested by these traps which

[1] *The Harvest of the Sea*, by James G. Bertram.

are then removed bodily and placed in other waters, where the eggs in due course hatch out and replenish the stock of fish. This rough-and-ready method is unfortunately of no avail for propagating trout, the eggs of which do not float.

A more striking example of Chinese ingenuity in fish hatching is mentioned by Bertram as 'a piscicultural novelty.' 'When the proper season for hatching arrives,' he says, 'they empty a hen's egg by means of a small aperture, sucking out the natural contents, and then, after substituting fish-spawn, close up the opening. The egg, thus manipulated, is placed for a few days under a hen.' With less than his usual thoroughness Bertram omits to describe the result of this charming method of incubation ; and one is left to speculate whether it will be a fowl or a flying-fish. It is not fair to condemn a new method without trial. But not having myself tried this Chinese novelty, I hesitate to recommend it to those of my readers who look for practical results in trout culture.

In the days of the Roman Emperors vast sums were spent by luxurious nobles on fish rearing in Italy. Huge reservoirs were constructed with channels connecting them with the sea. Lucullus is said to have supplied his house at Tusculum with salt-water fish in this way. Oyster farming was introduced

by Sergius Orata ; and the oysters of the Lucrine
Lake were held in high repute by the epicure.

That the monks practised fish rearing in ponds
attached to their monasteries is well known ; and the
distinction of being the first to conduct the artificial
hatching of fish-eggs in Europe has been attributed
to a monk of the fourteenth century.

To a German named Jacobi, however, belongs the
credit of having, more than a century ago, published
descriptions in scientific detail of his successful
experiments in the artificial spawning of fresh-water
fishes, including the trout. But the researches and
discoveries of Jacobi and of others who followed him,
though they attracted attention among the learned
biologists of the time, did not lead to the immediate
recognition of pisciculture as a practical science. It
was reserved for two French peasants, Remy and
Gehin, unaided by knowledge of previous or contemporary experiments, to rediscover, so to speak,
about fifty-six years ago, the art of fish hatching and
to demonstrate a few years later its practical utility as
a means of re-stocking rivers. The value of their discovery was soon recognised by the French Government. A grant of money was made, and the great
Piscicultural Institution at Hüningen in Alsace, near
Bâle, was thus in 1854 established by the erection of

suitable buildings under the direction of Monsieur Coste.

The system arranged at Hüningen by Monsieur Coste for the artificial propagation of fish has led some writers to regard him as 'the father of pisciculture.' The peculiarity of his system was the use of a grille or gridiron made of hollow rods of glass arranged closely together, the ends being fixed in a wooden frame. The eggs were laid on these grilles instead of being placed under gravel. The excellence of this plan is attested by the fact that it is still employed by some of the leading fish-culturists in Europe with the best results.

The tireless efforts of Frank Buckland in the cause of fish culture since 1862, and especially the publication of his fascinating little book 'Fish Hatching,' in 1863, stimulated many in this country to turn their attention to the artificial propagation of salmonidæ, and since that time numerous establishments for the cultivation of salmon and trout on a commercial scale have sprung up in different parts of the United Kingdom, some of which have met with conspicuous success. In his 'Natural History of British Fishes,' the preface to which, written at the end of 1880, is said to contain nearly, if not quite, the last words he ever wrote, Frank Buckland pressed upon his readers the claims of the art he had made so pre-eminently his own, saying that he regarded that book ' not merely as a book on ichthyology, but as an introductory guide to the great and important science of fish culture.' 'Throughout the country,' he says, 'there are streams as well as lakes and ponds suited for trout which are now troutless. Trout are becoming so important, not only as food, but also for trout fishing, that, wherever it is possible, I earnestly recommend that they should be bred artificially, and the waters thereby stocked.'

Since these words were written a change has come over the public mind. It is no longer necessary to urge the importance of pisciculture, and to explain the advantages of artificial stocking. This change is well illustrated by the words of the late Sir James Maitland, than whom there was no higher authority on trout culture in this, or perhaps any other country. Writing in 1892 he says, in the fourth edition of his valuable 'Pamphlet on Stocking,' 'The public, at least the angling portion thereof, is already very much up to date on the subject. The size and colour of eggs, the length and shape of yearlings or two-year-olds, are questions on which it holds a very decided opinion ; nor is it any longer necessary to point out

L.

the gain to be derived from stocking neglected pieces of water. The public wants to know what each pond, lake, or mile of stream should yield to the rod annually, and what species, kind, and description of fish will give the best sport.'

Later still, we have the weighty and disinterested testimony of Mr. Frederick M. Halford, by whom the idea that the natural reproduction alone in a river much fished is sufficient to keep up the stock, is ridiculed as 'an exploded theory' capable of being held only by such as would remain unimpressed by 'any amount of argument, or careful statistics of figures, or examples of the great benefit accruing to various fisheries from systematic stocking.' [1]

[1] *Making a Fishery* (1895).

CHAPTER II

REARING

WATER AND HATCHING-HOUSE

A SUITABLE water supply may be described as the first requisite of trout culture. Nothing can be done without this. On it everything depends. To secure a flow of suitable water must, therefore, be your first concern. Water taken from a brook or river is the best for rearing the fish, but for hatching the eggs use spring water if possible. A spot which yields both spring water and brook water gives you a great advantage. It need hardly be said that the water must, in either case, be free from all suspicion of pollution. Iron, lime, or sulphur, if present in great excess, unfits the water for trout breeding in any stage, but occasional muddiness is not a disadvantage in brook water for rearing the fish after the hatching period has passed.

Having found water suitable in quantity and purity, there still remains the vital question of supply.

Is it ample, and will it continue ample always, day
and night ? Will the flow of water, like Tennyson's
Brook, 'go on for ever'?

In this respect it should be able to stand the
severest test ; even after prolonged drought the stream
must still continue to flow in plenty. If it will do
this, well and good ; if not, then look elsewhere for
your water supply.

But this is not all. The opposite extreme must
be guarded against. It is not of course a disadvantage to have a greater supply of water than you
require to take, but there must be no risk whatever of
flooding ; the situation must be one in which you
will under no circumstances lose the power of control
and regulate your flow of water as you may think fit.
Nor must the water be liable to become overheated in
the summer through exposure to the sun. Beware of
water, the temperature of which is liable to rise above
70 degrees Fahrenheit. If you limit your operations to
hatching a small number of eggs and rearing a correspondingly small number of fry, all this can, with
suitable apparatus, be accomplished readily by means
of the ordinary domestic water supply of a town or
country house.

I recommend you to protect your spring or supply-
cistern with a closely fitting cover to exclude animals,

leaves and dirt, and to screen the water from the sun's rays.

Having secured a plentiful and unfailing supply of clean, cool water, beyond the reach of floods, your first difficulty is overcome. The next thing is to provide yourself with the necessary apparatus for hatching the eggs, and rearing the young fish.

Do not attempt to conduct your hatching operations in boxes placed out in the open air. From cold there is little to fear for eggs or the fry, but unevenness of temperature is very undesirable, and all risk of accidental disturbance has to be securely guarded against. Moreover the operator himself will require shelter during the hatching season, which occupies the winter months. A house or shed therefore of some kind should be brought into requisition.

Almost any kind of weather-proof structure can be made to answer your purpose, but a conservatory or greenhouse is not generally a suitable place for hatching and rearing operations, which require an even temperature. A hatching-room need not be an elaborate or expensive erection ; it should be easy of access from your house, and not far from the source from which your supply is taken.

The glare of the sun's rays is very injurious to the

eggs and also to the newly hatched fish. A dim light suits them best. The interior of the hatching-house should therefore in the daytime be shrouded in twilight. But you cannot work in the dark ; light therefore should be admitted as and when you require it through windows or skylights screened by movable blinds. These blinds should be kept closely drawn, especially on the south side, at all times except when you are actually at work.

The exclusion of the sun's rays helps to keep the temperature even, and to prevent the growth of a fungus which strong light has a tendency to foster. But there are worse enemies than the sun's rays to be shut out. Rats, mice, weasels, birds, &c., prowl and hover around bent on plunder ; therefore have a well-fitting door, and let this door be provided with lock and key, and on no account allow visitors to enter unattended. Smoking in your hatchery should be ' strictly prohibited.'

To secure the necessary fall the house should be so arranged or situated that the water may enter the room several feet above the floor. For this purpose the soil can, if necessary, be excavated, and the floor sunk to the required level ; or the building can be erected on lower ground. This will enable the hatching-boxes to be raised to such a height as to be

convenient for you to examine the eggs without much
stooping.

The supply-pipe leading the water into the hatchery
should be protected against frost. In the case of a
spring this may be done by laying the pipes underground. The receiving end of the pipe should be
covered with a cap of perforated zinc, to prevent the
possibility of any object alive or dead causing an
obstruction.

It is necessary that the water which passes over
the eggs should be clear and, as far as possible, free
from sediment which is injurious to the eggs. Good
spring water requires very little filtration, but if you
are dependent on a brook or river for your supply, the
water will have to be made to pass through a filter
before it enters the hatching-boxes ; and if the stream
is very muddy you will have to adopt the further precaution of allowing the water to clear itself in a settling-
tank.

Before entering the hatching-boxes the water
should be exposed to the air. This is specially
desirable where your supply comes straight from a
covered spring, and the water first sees daylight in the
hatching-boxes. The water is therefore passed into
a long open trough or tank placed across the upper
end of the hatching-boxes, and as high above them as

the level of the water outside will allow. To secure more complete exposure to the air I recommend that the water should be sprayed into the trough through a punctured pipe after the fashion of the sprinkler of a city water-cart.

The trough which I have just described is called the distributing tank, because its primary object is to distribute the water-supply in such a way that each hatching-box or range of boxes may receive its proper share of water. Outlets therefore are made in the distributing tank corresponding to the number of rows of boxes to be fed ; and each outlet is provided with a screw-tap. These screw-taps will enable you to regulate to a nicety the supply sent into each box. Pieces of india-rubber piping of· the required length serve to conduct the water from these taps into the hatching-boxes. Near the top of the distributing-tank there should be an overflow-pipe, and this should be kept always running.

THE HATCHING APPARATUS

The contrivances for incubating and hatching the ova are very numerous. Here, however, I propose to describe only one ; but this one I can thoroughly recommend, having myself employed it for many years with the most satisfactory results. Simple in con-

struction and easy of management, it is *par excellence*
the system for beginners.

It consists of a long deal box or trough, fitted as
shown by the accompanying illustration (fig. 8). All
the interior parts of it which will come in contact with
the water are carefully charred with a hot iron to
prevent the growth of fungus ; and this carbonising
process is repeated every year.

The size of your boxes will of course be regulated
by the scale on which your operations are to be conducted. An amateur making his first experiments in
trout culture may perhaps be content with a small
model 3 ft. 6 in. long, by 10 in. wide, by 6 in.
deep, with two compartments, as shown in the illustration (fig. 8). A hatching-box even of these modest
dimensions has often shown very creditable results
in the hands of a beginner. Of course you are not
limited to a single box. A series of them can easily
be arranged, one a few inches below the outlet
of another, in such a way that the water circulates
through every box. Near the top of the upper end of
the box a hole is bored, through which the supply-pipe
is introduced. Parallel with the upper end and two
inches distant from it is inserted a crosspiece, or
watertight partition, rising four inches in height from
the bottom which serves the purpose of a breakwater.

FIG. 8. Combined hatching and rearing box, Weston model.

A similar breakwater is inserted midway between this and the lower end of the box. Whatever the dimensions of your box are, the height of the breakwaters will always be the same.

A sheet of perforated zinc (No. 7 size), called a screen, is fixed in front of each breakwater in a slanting position, with its lower end up-stream, at an angle of 45 degrees with the bottom of the box. A similar screen is inserted in front of the lower end of the box. In fixing these screens care must be taken to leave sufficient space between them and the breakwater to admit your hand for cleaning purposes.

Unlike the breakwaters, the screens reach to the top of the box. They are very carefully fitted to the sides and bottom of it. The lower or outlet end of the box rises only four inches high (the same height as the breakwaters) and attached to the top of it is a zinc shoot to carry off the water. The illustration shows only two compartments. But the number is increased if the box is made larger. I need hardly say that a large hatching-box, requiring as it does more water than a small one, needs also a wider inlet.

Before beginning operations, raise the upper end of the box until the water stands in each compartment at a level two inches below the top of the upper

breakwater, when the water is running over the lower
breakwater, and prop it very securely in this position.
The water, released from the distributing tank by
turning the screw-tap, enters the hatching-box, and is
headed up by the first breakwater. Over this a
broad strip of zinc, laid evenly along the top, shoots
the water clear into the compartment below, creating
a miniature waterfall as wide as the box. The current
is thus distributed evenly, and the water thoroughly
aërated at the most important point in its course.

Passing through the first compartment, and
through the first screen, the current is arrested by the
second breakwater, which keeps the water above it at
a constant level. In the same way the head of water
in the last compartment is prevented from falling or
rising by the outlet-end or tail-board of the box,
and is ultimately discharged by the shoot attached
to that end. All the zinc parts of the apparatus,
whether fixed or movable, are varnished with Japan
black to prevent them from corroding.

Let us now turn our attention to the central point of
interest, the actual receptacle or nest in which the eggs
are to be ' nursed into life.' This is a form of what is
known among fish-culturists as a deep tray, so called
because in comparison with other forms its sides are
less shallow. The tray is made of perforated zinc (size

No. 7). It is rectangular in shape, and being made narrower by at least an inch than the hatching-trough, a sufficient space is left to allow the water to circulate through the side perforations.

Lowered into one of the compartments until the bottom of it is immersed to a uniform depth of two inches, the tray is held suspended in this position by two adjustable slings, each made of a single piece of stout wire bent to fit the tray. The ends of the

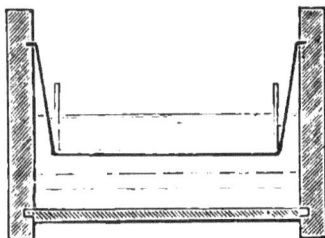

Fig. 9.

slings rise about two inches above the top of the tray, and, being bent outwards and pointed, are pressed into the sides of the hatching-box, where they are held fixed by the elasticity of the wire. The cross section (fig. 9) shows the tray slung in position by this simple but effective device.

The exposed side view given in the larger illustration (fig. 8) shows that the tray is so adjusted that the bottom of it lies parallel with the surface of the water, and not with the bottom of the hatching-box. Thus are all the eggs kept immersed to a uniform depth, while the water, the moving spirit of your apparatus, percolating through the tray at all points, circulates freely round each individual egg.

A similar tray is suspended in the same way in the second compartment. Each tray is furnished with a detachable cover made of blue calico stretched on a light framework of charred deal, which serves to keep the eggs in darkness.

Not the least important part of the box is a strong, closely fitting lid to prevent the incursion of marauders, such as rats and mice. The lid is divided longitudinally into two leaves hinged, one to each side of the trough, and opening independently. The hinges are so arranged as to allow each leaf, when opened and folded back, to hang vertically, flat against the side of the trough. Panels of perforated zinc let into each half of the lid afford the necessary ventilation.

This model hatching-box with fittings complete can be supplied at a cost of only forty shillings. In the early days of fish culture the eggs were laid in gravel in the hatching-boxes. The idea was to conform as far as possible to the natural conditions. The objections, however, to the use of gravel were many and serious. It was impossible to ensure cleanliness, or to keep the ova under close observation, or to remove dead eggs without disturbance. These and many other drawbacks have led to the total disappearance of gravel from modern hatching-boxes. It is of no use flogging a dead horse ; and it would be an ungrateful task, with the help of knowledge derived from later experience, to examine and dispose of the arguments by which those honoured pioneers and champions of fish culture, Frank Buckland and Francis Francis, used to defend with so much vigour the practice of incubating trout eggs in boiled gravel.

The glass grille system introduced by Monsieur Coste, to which I have before referred, has many able and experienced advocates who can point to results equal perhaps to those obtained by the use of any other system. Grilles are, however, undoubtedly more expensive than trays ; and the modification of the deep-tray system which I have described and recommended has, I think, distinct advantages on the ground of simplicity of construction, and the entire absence of difficulty in being worked even by a beginner.

The sides of the tray, standing as they do high and dry out of the water, afford a most convenient handle which can be grasped when necessary without wetting the operator's fingers.

Speaking of the deep-tray system, Livingston Stone says, 'The whole thing is so simple that nothing simpler that answers the purpose can be conceived. There is no complication of parts. There is nothing, in fact, to look after or move but the basket itself.' And he adds, ' one form or other of the tray system is undoubtedly destined to entirely supersede the old methods of hatching on glass grilles.' [1]

I should not myself venture to go quite so far as this, in anticipating what destiny or human invention may have in store for the fish-culturist of the future. In uttering the prediction I have quoted, Stone seems to have assumed that the American plan of pouring into a tray gallons of eggs until they were piled up ' 12 or 15 tiers deep' would, without doubt, be adopted universally.

At present, however, the pisciculturists of this country have not taken kindly to the American system, although we are assured, on the authority of Stone himself, that the ova ' suffer no injury whatever from being so piled up ; one explanation of this being that the water all the time forcing its way up through the eggs, loosens them so that they do not feel the

[1] *The Domesticated Trout*, by Livingston Stone.

weight of those above them, while at the same time it reaches every egg, and furnishes a fresh supply of air to them all.'

COLLECTION OF THE OVA

Everything in the hatchery should have been tested and got into thorough working order before the autumn ; indeed, it is a good plan to keep the water running through the hatching-boxes for several weeks before the trays receive the ova. In this way you will be enabled to detect any weak spot in your arrangements, and to make sure that everything is properly and permanently adjusted. Remember that when once your eggs are laid down, any interruption involving stoppage of the water supply will entail damage if not disaster.

The English trout is a winter spawner. But the exact period at which the fish begin to spawn in any given locality must be ascertained by observation and enquiry on the spot. Much will depend on the climate, the season and the condition of the water. But at some time between the latter part of October and the beginning of February the trout will be observed working their way in pairs into the small tributary streams, and into those parts of the main

M

river where the water is shallow, the current rapid, and the bottom gravelly.

The spawning season has then come. The fish now begin to build their nests. This they do by fanning up the gravel with their tails until a trench is hollowed out, and a mound formed immediately below it. These nests, called 'redds,' can be easily recognised by the contrast of colour presented by the under surface of the newly turned stones. Into the trench the female enters, followed by the male, and here the eggs (the hard roe) deposited by the female and fertilised by the milt (the soft roe) of the male are covered over with gravel by the parent fish.

I have said that the eggs are fertilised. But as a matter of fact a very large proportion of them escape coming in contact with the milt and so perish, or rather fail to come to life, for want of being fecundated. The magnitude of the waste from this cause alone has been estimated at more than one half of the entire amount of spawn cast, and it is in remedying this failure that the fish-culturist, coming to the aid of Nature, achieves, as we shall presently see, the first and perhaps the most conspicuous of all his successes.

Having ascertained that the fish are on the point of spawning you will proceed to capture them with nets. This is generally a very chilly and laborious task. Nor is it free from the element of uncertainty, for the spawning period in a given locality lasts only a short time, and unless you have been observing the spawners closely you may be too early, or, worse still, too late to accomplish your object, which is to catch the gravid fish in the nick of time, viz. on the eve of spawning.

Having succeeded in this you will place the mature females in one tub or bath and the mature males in another. The sexes are easily distinguished at this season ; but it is experience alone which will teach you to recognise, at a glance, with unerring certainty, which of the fish are forward enough. In the female there is a fulness and a looseness which is unmistakable, showing that the ripe eggs have, to use Buckland's phrase, 'detached themselves from the membrane, and lie quite loose in the cavity of the abdomen.' I do not know that I can better describe this than by quoting the words of the veteran American fish-culturist Seth Green : 'The ripe female, by an expert, can be told by the touch. When in perfect condition, the belly of the fish feels soft and flabby, and if she were held up by the head her spawn would settle downwards, while with the unripe female the spawn remains stationary in place and feels to the touch hard and bunchy, like shot.'

The eggs of the mature female have become disconnected, and will shell out like peas at the slightest pressure of the hand. The maturity of the male fish is also easily recognised, as the milt will begin to flow almost spontaneously. It is just as well not to attempt to manipulate a fish until it has been allowed to kick in the landing-net for a few minutes and has become quiet.

There are many methods of taking the eggs. Each man ultimately hits on a plan most convenient to himself. A satisfactory method of manipulation is as follows : Using a piece of flannel or cloth to prevent slipping, you seize a female by the tail with your left hand, and thrust her head between your right wrist and your chest. In dealing with a big fish the head will go under your arm. The head is thus steadied, the fish's back being against your chest, and the belly turned outwards. Now close the fingers of your empty right hand, and slowly extend them again in such a way that the tips of your fingers and the ball of your thumb pass gently along the whole length of the abdomen, beginning at the pectoral fins. Repeat the process until all the eggs have come out, taking care that your finger tips always

follow the eggs and do not run in front of them.
Simultaneously with the movement of your right hand
the fish's tail is slightly bent back with your left
hand. Avoid all pressure on the lower part of the
abdomen ; and on no account use force to press out
the spawn. Your efforts should be directed towards
making each fish, as far as possible, spawn itself.
Clumsy handling is often responsible for an injury
to the eyeball of a spawner, resulting at a later stage
in blindness.

A common glazed earthenware milk pan held
close to the fish by an attendant receives the eggs.
The pan should be clean, and should contain no
water except the few drops which may fall from the
fish. Sometimes a trout, although mature, will hold
back its spawn. But in any case, if the eggs refuse
to exude readily take another fish, and try the first
one again later on.

The amateur will probably not be very neat-
handed at first. By unskilful manipulation he may
cause the fish to struggle, and may even allow it to
slip from his grasp into the spawning pan, thereby
destroying a quantity of eggs. But practice makes
perfect, and you will soon acquire the knack of strip-
ping the fish with such ease and rapidity that trouble
from struggling will be a rare occurrence.

The ova having been collected in the spawning-pan, should be at once fertilised. In order that there may be no unnecessary delay, you should not begin to take the eggs unless you have ready at hand a sufficient number of males.

The fertilisation of the ova is accomplished by what is known as the Russian or dry method.[1] Without adding any water to the eggs, take a ripe male, and dealing with it in very much the same way as you treated the female, allow the milt to flow over the eggs. Stir both milt and ova together gently with the fingers, taking care to keep the finger tips on the bottom of the pan so as not to squeeze the eggs, which at this stage are soft and adhesive. Then add one or two handfuls of water, about a wineglassful in all, and just enough to immerse the eggs.

[1] So called from having been first discovered by M. Vrasski, a Russian. The discovery was the result of a series of experiments made in 1856. But it was not until 1871, when the discovery became known to American fish-culturists, that the old method of taking the eggs in water and then adding milt to the water began to be discontinued. The experiment of comparing the results obtained from both methods would be very startling to one who did not know what to expect. Sixty or seventy per cent. was a good average percentage of impregnations under the old method. But under the new system only one or two eggs in every hundred are likely to escape being vitalised. The reader will not, therefore, be surprised to learn that the older or 'more natural' method has now become obsolete.

Covered with a cloth the vessel is now left undisturbed for about 25 minutes, or even less, according to temperature. The colder the water the longer
the eggs must be left. At the end of the time the
ova will have separated themselves and become
hardened. Now pour in clean water, and having
stirred the eggs with the hand, pour off the water
again. Continue this washing process until all the
effete milt has disappeared, and the water leaving
the ova is as clean as that poured on them. Transferred into a collecting can filled with water, the
eggs are then ready to be taken to the hatching-
box.

As soon as the fish have been stripped they
should be set at liberty again ; and unless they have
been very clumsily dealt with, they are none the worse
for their experiences.

For the instruction of my readers I have indicated
the proper method of taking and vitalising the eggs
of wild trout. But to a beginner, especially if his
hopes are fixed on ultimate results and 'percentages,'
my advice is to avoid the difficulties, risks, and uncertainty, not to mention the personal discomfort,
involved in collecting the ova. Let him take a short
cut, by purchasing eggs at a later stage of development, known as ' fully-eyed ova,' from a leading fish-

culturist, who undertakes to supply them at a reasonable price.

Produced from large, healthy, selected fish which have been reared and conditioned for the purpose, such ova are much larger, and should give you a far better result than eggs taken indiscriminately from wild fish by inexperienced hands.

Great differences are observable in the colour, the size, and the vitality of ova taken from different trout. The colour follows that of the female fish's flesh, and the hue of the flesh being determined by diet, it follows that the colour of the ova is attributable to the spawner's feeding.

In batches of eggs from different localities all variations of tint are exhibited, from 'splendid coral red' to a pale yellow, some again being 'almost white as peas,' yet all healthy ova.

Eggs, the produce of large well-conditioned females of 3, 4, and 5 lbs., are much superior to the ova of small hill-burn trout. In the first place, the eggs are larger—that is to say, they run fewer to the gallon.

Sir James Maitland's figures are : for first-class trout, 28,000 eggs per gallon, ordinary trout, 40,000, well-bred trout, 32,000, ordinary wild lake trout, 40,000, wild hill-burn trout, 50,000.

Trout rearing has often been compared with sheep breeding. But here is a point in which the two sciences differ. For his ewes to bear only one lamb apiece is not an advantage to the sheep farmer, even though each lamb be one of singular promise. The more lambs you get from a given number of ewes the better. It is not so with trout. What the pisciculturist wants is big eggs. He wants quality rather than quantity. A big egg has usually a stronger vitality, and produces a larger and stronger fish. I may mention here that the very best results are obtained when the ova of selected breeders have been fertilised by the use of wild males.

The number of eggs produced by the salmonidæ is very small compared with the number produced by other fishes. One thousand to each pound of its weight is considered an average number of eggs to be deposited by a spawning trout, which does not weigh less than a pound. A salmon's average is even less, being about 800 or 900 for every pound of the fish's weight.

According to Frank Buckland's table a pike of 4½ lbs. has 42,840 eggs, a perch of ½ lb. 20,592, and a roach of ¾ lb. no less than 480,480 ; while in a sole of 1 lb. the number given is 134,466, and in a cod of 20 lbs. the total reaches to nearly five millions. When the significance of these figures comes to be considered it will be seen that the trout stands at a disadvantage in the matter of reproduction.

But if birds of many different kinds with only four or five eggs apiece continue plentiful, how is it, one may ask, that trout with upwards of a thousand eggs each are not able to keep up their numbers without man's assistance? In other words, where does the utility of the fish-culturist's art come in? This is a very pertinent question. Therefore on the journey home with our precious burden of fertilised trout ova, let us pause to consider what would have befallen them had they been deposited by the parent fish in the gravelly recesses of the natural spawning-bed.

As we have already seen, a large proportion would have perished for want of being vitalised. But what of the remainder? Many being ineffectually caught by the gravel are washed away by the current. Numbers too are buried alive under heaps of mud and débris brought down by floods. The waters of the shallow tributaries sometimes recede leaving the nests high and dry, with the inevitable result. In salmon rivers a very severe frost has been known to affect the spawning-beds and to destroy the whole of one year's crop of eggs, and I have little doubt that to some extent trout eggs occasionally meet with wholesale destruction from the same cause.

But the effect of these agencies taken together is as dust in the balance when weighed against the devastation attributable to various living enemies. 'There is not a living creature,' said Mr. Francis Francis, 'which inhabits the waters, which does not prey more or less on trout ova.' Without concurring in this very comprehensive verdict of guilty, I will go as far in that direction as to say, that you will be safe in reversing two popular English law-maxims by assuming that every living inhabitant of the waters is guilty of egg-slaughter until he is proved innocent, and by never giving the accused the benefit of the doubt. It is on this working hypothesis that artificial incubation is conducted, and the trout-breeder who hesitates to exclude animals of every sort and kind from his hatching-boxes is lost.

The trout themselves are incorrigible rogues in respect to the ova of their own species, which they consume in large quantities. And there is another way in which the trout themselves cause damage. It frequently happens, especially where the spawning-beds are not sufficiently extensive, that a later set of spawning fish come up and play havoc among the eggs by turning up the nests made by the earlier fish.

Buckland mentions a case in which trout were seen in the act of devouring the eggs, 'with their noses grubbing in the nests and their tails projecting out of the water like so many sharks' fins out at sea.' Ducks, geese, swans, moorhens, dabchicks, and many other kinds of feathered fowl delight to gorge themselves with the contents of the spawning-beds. Eels, roach, dace, chub, and other coarse fish attend in great numbers on the spawning trout and regard their eggs as a dainty dish. Freshwater shrimps are very destructive to trout eggs. So too are the larvæ of the Mayfly, the dragon-fly and many other water insects.

You will now have no difficulty in realising that, of all the useful functions performed by the processes of trout culture, one of the most important is the protection of the eggs under conditions which place them beyond the reach of the many dangers to which in the natural state they are continually exposed. Nor will you find it difficult to credit the statement so often made, that out of every thousand trout eggs deposited naturally, only one ever becomes a mature fish.

It is a very remarkable fact that during the earliest stage of development—viz. within forty-eight hours after having been vitalised—the eggs are capable of bearing a considerable amount of disturbance and

movement. But the germ of the embryo having risen
and 'furrowing' begun, the ova become extremely
sensitive to the effect of vibration or concussion,
until a later stage (to be treated of in the next chapter)
when they again become tough.

The hardy character of the eggs in their earliest
stage enables them to endure a journey of considerable length, immediately after being taken ; they then
travel well in still water, which would prove fatal to
ova in a later stage of development.

But hardy as the eggs are at first, any violent
shaking or concussion is even at that stage calculated
to produce numerous dead ova, or an undue proportion of weakly and deformed fry.

HATCHING

With as little delay as possible you will take the
eggs to the hatching-box and carefully empty them
out into the water in the trays. Do not remove the
trays from the boxes for this purpose. By gently agi-
tating the trays under water the eggs will distribute
themselves evenly. If necessary a feather can be
used to move the eggs under water, but the less they
are touched the better.

Having adjusted the calico covers on the hatching-
trays, draw the window blinds, lock the door of the

hatching-house, and leave the running water to perform the work of incubation. If your hatchery is thoroughly secure against intruders, it is not necessary to close the lids of the boxes. The tray-covers will afford sufficient protection against a stray mouse.

Your labours during the hatching stage will not be very arduous. You must, however, inspect the boxes daily to make sure that the water is continuing to flow properly, and to remove any dead eggs. A bad egg is easily recognised by its milk-white colour, and unless it is quickly taken out, the ova which lie in contact with it will decay, and, if the mischief is not arrested, a form of fungus called *byssus* will eventually appear and spread devastation far and wide. The injurious effect of a rotting trout egg in a hatching-box has been aptly compared to the mischief which one decaying peach will communicate to the good fruit which lie close to it in the same basket.

In the old days, when the ova lay concealed from view under gravel, it was very difficult to ensure the removal of every dead egg. *Byssus* was then a real danger, but with a modern system of incubation on trays or grilles none of the eggs can long escape observation; nor should the scourge of *byssus* ever afflict you except as a penalty for very serious neglect.

Byssus must not be confounded with *saprolegnia,* another and a more dangerous fungoid growth which attacks fish as well as eggs. It is to this more formidable growth that the fish-culturist is always understood to refer when he speaks of 'fungus.' The grand preventives of saprolegnia are charred wood, cleanliness and darkness.

To enable you to remove the dead eggs without disturbing or even touching the others you must use an egg-extractor. The instrument I recommend is a modification of the pattern recommended to me some years ago by Mr. Silk, the then manager of the Marquess of Exeter's fishery.

It consists of a tube furnished with an india-rubber suction bulb at one end (fig. 10). The tube in the

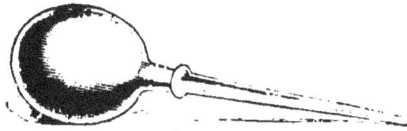

Fig. 10.

original pattern was of wood; but it is an improvement to have it made of glass.

To use this instrument you squeeze the bulb and having placed the nozzle of the tube against the egg, immediately relax your grasp slightly. The egg is then withdrawn adhering by suction to the end of the tube, the bore of which is not large enough to admit of the egg being 'swallowed.' Collect these dead eggs and throw them away, not on the floor of the hatchery, but at a safe distance from it.

Be on your guard against sediment. Even the best of water contains fine particles of dust in suspension. These particles settle on the ova and can be clearly detected by examination of an egg through

FIG. 11.

a magnifying lens. It is here that a dipping tube (fig. 11) is serviceable. It consists of a bent glass tube of large bore (¾ inch) with enlarged elbow, and is provided, like the egg-extractor, with a rubber suction-bulb.

To remove sediment, you simply lift out the hatching-tray, and thoroughly sprinkle the eggs with water by means of a small watering-can fitted with a very fine rose. This watering of the eggs has I think a good effect, even if there is no sediment to be removed.

With good water taken underground from a strong

deep spring there should be very little trouble from
sediment ; but if your supply for the hatching-boxes
is taken from a river or burn, constant attention will
undoubtedly have to be given to the filtering arrangements, otherwise damage will ensue.

A suitable kind of filter consists of a wooden box
containing three or four screens made by tightly
stretching flannel on wooden frames. These screens
fit in grooves cut to receive them in the sides of
the filter-box at intervals. The interior of the box
is charred in every part. An extra set of screens
accompanies each box ; so that when one set has
become choked, another may be substituted while
the screens first used are being dried and cleaned.

The process of hatching trout-eggs with proper
modern apparatus does not involve heavy labour.
Nor does it present difficulty or demand special skill.
Indeed it has been described by Livingston Stone as
'the very simplest and surest of all the branches of
trout breeding.' At the same time patience is required.
The stages of development of the embryo in the egg
occupy a considerable time. The length of time is
mainly dependent on the temperature of the water.
With an average temperature of 48 degrees Fahrenheit
the eggs will hatch in about fifty-three days from the
time when first laid down.

N

This period will be longer if the water is colder, and shorter if it is warmer. One degree Fahrenheit makes a difference of about five days in the time. It is not an advantage that the eggs should hatch out quickly. Your object is to produce firm-bodied healthy fry, and this is not to be done by raising the temperature of the water, or by admitting strong light, and so unduly hurrying the embryo through its various stages of development.

The space at my disposal does not allow me to describe in detail the successive changes which, during the period of incubation, go on inside the transparent, horny shell of the ovum, intensely interesting as these changes are to the observer. But I may mention that when about a third of the hatching period has passed two black spots will become plainly visible in each egg. These spots are destined to become the eyes of the fish, and the eggs have now reached the stage at which they can, when properly packed, be moved even to very long distances without injury. They are known as 'eyed ova,' and it is at this stage of maturity that the eggs will arrive which I have advised you to purchase from a leading fish-culturist.

For packing, small shallow trays are used, made of muslin, stretched on light wooden frames. These trays with a layer of eggs on each are piled one upon another in a box, and in this way the ova will reach their destination uninjured after a journey of several days.

In packing eggs for shipment to the antipodes, layers of live, damp moss are inserted between each tray, and the later stages of development of the embryo are lengthened, and the date of hatching deferred, by the use of ice. The live moss, by absorbing the carbonic acid given off by the eyed ova, enables them to take in fresh oxygen, without which they would be suffocated, just as they would, at this stage, be drowned in a few hours if put in a small bottle of water and kept still.

The toughness and vitality of the ova of the Salmonidæ after reaching the eyed stage are very remarkable, and present a striking contrast to their extreme delicacy throughout the whole of the preceding stages, except during the first few hours after being taken. From day to day after the appearance of the eyes, the colour of the embryo will grow darker and its outlines more distinct, until at last you can see it twitching in the shell as if struggling for liberty. 'The plot thickens,' and you are on the tiptoe of expectation. The hatching time is now close at hand, and you may expect any day to be rewarded

N 2

for your patient care by finding in the trays 'something new and strange.'

What happens at the exciting moment of hatching has been described by Frank Buckland. 'If you have luck,' he says, 'you may happen to be gazing on a particular egg, when on a sudden you will see it split in twain at the part corresponding to the back of the fish ; you will then see a tiny head, with black eyes, and a long tail pop out, and you will see the new-born creature give several convulsive shudders in his attempts to quit himself of the now useless egg-shell.'

With large, strong eggs, properly incubated, there should be little loss in the actual hatching ; but if the embryos are puny, many of them will not have strength to disengage themselves from the egg-case, and a heavy bill of mortality will result.

Instead of laying fully eyed ova in hatching-boxes, they can be sown in an artificial ova bed formed in a stream connected with the water which you desire to stock. The stream should be kept completely under control, and the supply of water regulated by a sluice. The bed of the trench is of sifted gravel, and in this gravel the eyed eggs are planted just before they are expected to hatch out. Of course the bed must be screened from the sun's rays, and most carefully protected from the attentions of ' Fur, Feather, and Fin,' also from Jack Frost, and (need I add?) sewage.

An arrangement of perforated zinc screens at the inlet and the outlet, on the principle of a hatching-box, should be provided, and of course there must be a cover.

A specially constructed wooden ova-bed can sometimes be used with advantage. As soon as the alevins have begun to feed, the outlet screen is removed and they are allowed to work their way down the stream at their leisure into the water which they are intended to replenish.

Under such a system as this the percentage of loss, which is of necessity very heavy, is anticipated and to some extent discounted by the very large quantity of eggs laid down. It is, however, only under exceptional circumstances that the plan of sowing eyed ova in artificial redds can be adopted with advantage and economy. And you may well pause to take advice before incurring any heavy outlay about artificial ova beds.

THE ALEVIN STAGE

If you have been expecting to see a completely developed fish emerge from the shells, you are now disappointed. The new-born creatures are in the intermediate tadpole stage, in which they are known as 'alevins.' Attached to them underneath is a kind of bladder nearly as large as their bodies, known as the umbilical vesicle or yolk sac.

The American fish-culturist, Seth Green, has aptly described it as 'Nature's feeding-bottle.' It contains nourishment in the shape of oil globules, which are alone sufficient to support the infant until it is able to eat.

The accompanying graphic diagrams (fig. 12), taken

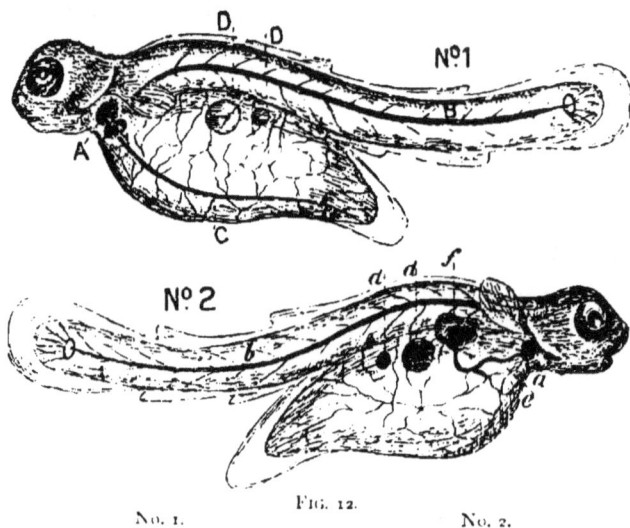

FIG. 12.

No. 1.

THE LEFT SIDE OF THE FISH.

A The heart.
B Main trunk artery.
c Large vein.
D Oil globules, only two of which are introduced for sake of distinctness.

No. 2.

THE RIGHT SIDE OF THE FISH

a The heart.
b Main trunk artery.
c The liver.
d Oil globules.
e Large vein from liver to heart.
f Branch from main artery to liver.

from Frank Buckland's 'Natural History of British
Fishes,' represent a newly hatched salmon. But they
convey also a very good idea of the appearance of
his near relative the trout at the same tender age.
For among the infants of the Salmonidæ group
there is a strong family likeness, although the baby
trout is, as you would expect him to be, smaller
than the king of fishes.

Examine one of the alevins under a microscope,
and you will be rewarded by a spectacle which has
been described by Buckland as 'one of the most
beautiful sights ever beheld by human eyes,' including
a complete demonstration of the circulation of the
blood.

The first impulse of the alevins seems to show
itself in eagerness to seek concealment and to avoid
light. Fitful and spasmodic are their earliest efforts
in locomotion. Encumbered by the huge yolk sac,
they wriggle about in a clumsy and aimless way, and
then sink to the bottom of the tray and lie panting
on their sides.

When you have feasted your eyes on the wonderful
transformation scene, replace the tray covers and
leave the hatching process to continue in darkness
undisturbed. On examining the trays next morning,
you will find that most of the eggs in it have hatched

out, and that there is a grand litter of empty shells.
These you will proceed to remove with the egg
extricator (fig. 10).

Soon the alevins will be seen to crowd together
in dense throngs about the sides and corners of the
trays where there is most shade to be had. This
'packing' of the alevins is a healthy sign. Closely
observed, their pectoral fins will be seen to be all
working at a tremendous pace ; and it is in this way
that each individual alevin assists in the work of
respiration not only for itself but also for the rest of
the shoal. In fact, as Sir James Maitland put it,
'the mass is a gigantic co-operative breathing association.'

Gradually the yolk sac will grow smaller, becoming
incorporated in the body of the fish. The length of
time occupied by this process is regulated chiefly
by the temperature of the water. From four to six
weeks is a common time to elapse before the little
fellows begin to feed, which, if they are strong,
should be shortly before the final disappearance of
the sac.

During the period of alevinage your labours will
again be light. Snugly secure the alevins remain
cribbed in the hatching-tray, the sides of which are
high enough to prevent them from jumping out.

Here they require little attention and no food; but you will inspect the trays daily and remove any corpse you may find. Take care to leave the calico covers on the trays and to keep the blinds of the hatching-house drawn, as sunlight is no less hurtful to the alevins than it was injurious to the eggs.

Scrupulous cleanliness is one of the first principles of fish culture in all its stages, and a great advantage of a hatching-box of the pattern I have recommended to you is that it is easily cleaned. I do not mean to imply that much cleaning work will be required at this stage; but even before feeding operations have begun a certain amount of impurity will

Fig. 13.

be found to collect on the sides and bottom of the spaces immediately below the screens. This you will detach with a common saucepan-brush (fig. 13) and draw off with a piece of ¾-inch-bore rubber piping about four or five feet long, used as a syphon.

In your daily battles on behalf of cleanliness at this stage, but more especially later, there must be no truce. Your armoury therefore must not be ill provided with effective weapons for dealing with the forces of uncleanliness. A more serviceable weapon you will hardly find than the rubber syphon. At its touch the daily gatherings of the enemy will disappear

and vanish, and you can close the boxes without fear
of an attack from fungus.

'There is no word,' says Livingston Stone, 'in the
fish-breeder's vocabulary, that is so associated with
loss and devastation as the word fungus. There is
nothing with which he has to deal that is so insi-
dious and deadly;' there must therefore be no
thought of *cure* for fungus. Absolute prevention
must be your policy. With boxes well charred, however, and with close attention to cleanliness, fungus
should never trouble you, unless you are subject to
nightmare.

Shortly before the last vestiges of the yolk sac are
absorbed the packed shoals of alevins begin to leave
the corners of the tray, to rise in the water and to
distribute themselves more evenly. All desire for
co-operation has left them, and henceforth their life,
unlike that of the minnow or the gudgeon, will be
individual rather than gregarious. Some of the alevins
will now be observed poising themselves in the water
and darting at particles which float past them, as if
making attempts at feeding.

As soon as the little occupants of the tray have begun
to behave themselves in this fashion, they will require
wider scope and a greater depth of water. The time
in fact has come when they should be allowed to have

the range of the whole compartment unrestricted by
the narrower confines of the tray.

But before this change is made, it is very desirable
that the floor and sides of the box should be
thoroughly cleaned in every part ; for at this delicate
stage of the alevins' existence, the possibility of contact with anything that is not perfectly sweet and clean
cannot be too carefully guarded against. The tray is
accordingly lifted out and suspended for a few moments in another box close at hand, through which a
stream of water is running. The compartment just
vacated can now be thoroughly cleaned with saucepan-
brush and syphon, the wire tray-supports having been
first removed.

Cleaning operations over, the tray is lifted back
and the tiny tenants evicted in a summary fashion by
the simple process of turning their crib gently upside
down and removing it altogether. The tray is now
laid aside with its calico cover and wire supports ready
for use again next season.

One of the advantages claimed for the deep tray is
that, unlike the grille, it does not allow the alevins to
drop through into the box beneath and disappear
from view one by one as the eggs hatch out, but keeps
them all in the tray under your immediate observation ;
and further, that when the time comes for the alevins

to leave the tray they have the distinct advantage of beginning their feeding life on a floor which has been thoroughly cleaned *immediately beforehand*.

The nursery in which the alevins now find themselves is certainly more useful than ornamental. On referring to the illustration (fig. 8) you will see that the floor and side walls of the room are formed by the charred bottom and sides of the hatching-box. The wall at the upper end is the breakwater, over the top of which the water enters, while the lower wall consists of the sloping screen of perforated zinc. For their ceiling the alevins now have the lids of the box instead of the blue calico cover of the tray which they have just vacated. Their new apartment is of course more spacious than the tray; but there is the same conspicuous absence of furniture.

REARING THE FRY

When once the alevins have begun to feed, the process of absorption of the yolk sac will be rapidly completed. Their digestive organs are now developed, and each tiny creature has become a fully formed fish. The period of alevinage has passed, and the little creatures are now entitled to be called by the more familiar name of 'fry.' Your labours have ceased to be nominal, and you must now put your shoulder to the wheel.

It is of great importance to get the fry to feed as soon as possible. By throwing in very small particles of food, and carefully observing the result, you will soon ascertain what progress is being made. The plan of turning out very young fry into ponds and streams immediately after the absorption of the umbilical sac has not found general favour amongst fish-culturists. The habit of feeding not having been acquired, the loss of a large percentage from starvation is inevitable. An eminent authority on fish culture has admitted that in transferring very young fry straight from the hatching-boxes to hundred-feet protected ponds he has allowed for a loss of nearly seventy per cent. from starvation, even though the fry were regularly fed by experienced assistants.[1]

The method which I favour is that of extending to the young fry the protection of the boxes for several weeks—viz. until they are larger and stronger, and have grown thoroughly accustomed to feeding. But especially do I recommend the use of the rearing-boxes where you intend afterwards to grow the trout

[1] *History of Howietoun*, by Sir J. Ramsay Gibson Maitland, Bart.

in rearing-ponds, and ultimately to turn out the fish
as yearlings.

For the first day or two after leaving the hatching-
trays the fry will hardly require to be fed more than
twice a day. But after that the feeding must be done
at least six or seven times a day for two months.

An amateur, if single-handed, is likely to find this
unremitting attention monotonous and irksome; and
I should not recommend him to disdain to call in
the occasional help of some reliable friend or assistant
to share his labours during the first seven or eight
weeks of the feeding stage.

Opinions differ as to what is the best kind of food
for trout-fry. The roe of soles, haddock, mackerel,
and other fish has been recommended and been used
with success. Various kinds of specially prepared
paste, biscuits and meal are in use, made of such
ingredients as shrimps, mussels, yolks of eggs, arrow-
root, and fillet of beef. Bullock's liver, baked and
finely grated, is not a bad food, and a mixture of curd
and liver has also been thought suitable by some
competent fish culturists. Raw sheep's liver, however,
is the diet which, on the whole, I have found most
satisfactory. The preparation of the liver is a work of
great importance. It is first cut up in a chopping-
machine. Currie's American meat and vegetable

chopper is the machine which I have found most serviceable for this purpose. The pieces of liver are then 'screened' by being pressed through a sieve. This sieve is a very simple contrivance. It consists of a small wooden box, the bottom of which has been knocked out and the top replaced by a piece of perforated zinc securely fixed by nails. At first the mesh of the zinc must be very fine (size No. 5).

After having been separated from the skin by scraping, the liver is rubbed through the perforations of the zinc with an ordinary iron kitchen spoon, and is collected in a small pail placed under the screen (fig. 14). So finely is the liver separated by this pro-

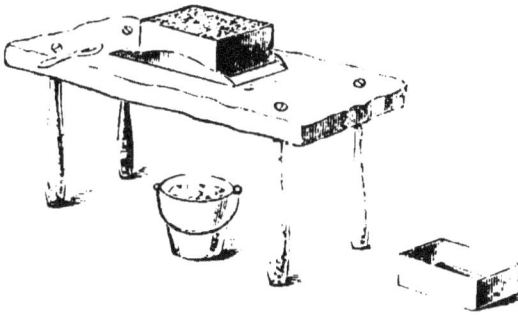

FIG. 14.

cess that it is reduced to the consistency of a thin paste. A little of it is now put into a 'feeder,' a utensil which consists of a cylindrical vessel made of a piece of finely perforated zinc of the same mesh as the sieve, soldered round a circular base of plain zinc (fig. 15). To use the feeder you dip it in the water in the rearing-boxes and shake it gently. A cloud of finely separated particles will be disseminated in the water, while any larger fragments will remain behind in the cylinder.

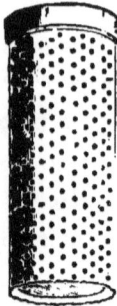

Fig. 15.

Healthy fry will eagerly snap up and devour the minute mouthfuls which float past them ; and, with regular and careful feeding the greedy little fellows will soon learn to recognise the feeding-spoon and, like Oliver Twist, to boldly 'send up their plates for more.' The appetites of healthy fry are indeed prodigious, and unless they are regularly and frequently fed they will take to nibbling each other's tails and fins—a most undesirable habit.

There is not very much fear of your feeding them too often. And with the constant cleaning which the boxes should receive, the unconsumed particles of food will have no time to decompose and con- taminate the water, or even to render it sickly and unwholesome.

Of course you must not discharge among the fry more than a small quantity of food at one time, otherwise some of it will sink to the bottom, and be

wasted. Even well-bred fry cannot be taught to
resist the temptation to seize the very largest piece
they can see and bolt it whole ; if, therefore, you
would avoid loss from death by choking, take care
not to allow the fish to have pieces of food too large
for their mouths.

In a small model hatching-box of the size I have
mentioned, you would perhaps be able to keep two or
three hundred fry for about a fortnight after they had
begun to feed.   At the end of this time it would be
necessary to thin them out in order to avoid the ill-
effects of overcrowding.

It is a good plan, therefore, to have ready at hand
extra rearing-boxes, made on precisely the same plan
as the box which you have been using, only larger.
They may be placed out of doors, and it is better
that they should be supplied with brook-water,
although excellent results have been obtained by
using the spring-water of the hatching-house.

An example of the successful rearing of fry to an
advanced stage indoors in a great city with water
supplied from the company's mains will be remembered by visitors to the Imperial Institute of 1897,
who may have seen there the exhibit from the Weston
Fishery.   In six weeks the fish had grown to two
inches in the boxes.

O

By making use of a distributing trough you
can have several rearing-boxes compactly arranged
side by side, and if the natural fall is insufficient, you
can, by excavating the ground, create the required fall
for the water, and be enabled to raise the boxes on
trestles or blocks of wood to such a height as will
make it unnecessary for the operator to stoop to his
work.

The side-by-side arrangement of the boxes is often
the only one which the height of your available head
of water will permit ; moreover, it economises space
and also saves time in feeding ; but, most important
of all, it enables each box to receive a supply of
water which has not immediately before been deprived
of much of its oxygen by passing through the gills of
fry in other boxes above. Very great care will have
to be taken to prevent the possibility of the water
supply being interrupted.

The task of transporting the fry from one nursery to another must be accomplished without removing the delicate little creatures from their native
element, an easy matter if only we call in the services of our old and trusty friend the rubber syphon.
In a trice it will suck out as many as you wish into a
fry-carrier a vessel provided with a panel of perforated
zinc let in near the rim which allows the surplus water

to escape but not the fish. During this operation the lower end of the syphon should be kept immersed.

It is a quaint sight to see the unwilling fry disappearing (always tail first) up the syphon and excitedly reappearing, again tail first, in the vessel below, none the worse after their first journey at railway speed through the unfamiliar dark tunnel.

In the outdoor rearing-boxes feeding will be continued as before in the hatching-house. The mouth of each fish, however, is daily growing larger and its digestion stronger. Soon, therefore, the very fine zinc feeding cylinder should be changed for one with larger perforations (known as No. 6), which will allow more substantial morsels to pass through. The zinc of the box sieve will, of course, undergo a corresponding change. 'Shade,' says Livingston Stone, 'never killed a trout yet, young or old; sunlight has killed a great many.' For this reason keep the lids of the rearing-boxes at first always closed on both sides when you are not attending to them. Gradually, however, the fry should be accustomed to more light by way of preparation for the time when they will be turned out; and here it is that you will realise the advantage of having the lid of each rearing-box divided into two independent halves; for by opening only that half which is farthest from the sun you can

admit light without depriving the fry of the benefit
of the sheltering shade afforded by the other half
which remains closed. At night both sides of the
lid are always closed, and a padlock is an excellent
safeguard if there is any fear of 'the fingers of
meddling boys.' During periods of severe frost it
may be necessary to protect all the boxes by covering
them with planks. The rearing boxes will want daily
cleaning with syphon and brush. Some kinds of food
cause more deposit than others, but with so many
mouths to feed in such a small place, unless your
cleaning work is done every morning with regularity
and thoroughness, the result will be tainted and unwholesome water, no matter what food you may use.
When this has happened a heavy penalty is likely to
await you for your failure to ensure cleanliness. You
will have left open the gates of your citadel to the
great enemy fungus, or to some foe even more insi-
dious and almost equally destructive.

A distressing feature of some of the worst maladies
of the young trout is that the effect of them does not
always make itself manifest immediately; and you
may be living in a fool's paradise, all unaware of the
presence of the enemy in your midst. Particularly so
is it with certain affections which attack the gills, the
fishes' lungs. The gill-covers conceal the ravages of

the disease in its early stages, and you may even remain in ignorance of the cause of the mischief until numerous deaths have led you to hold a post-mortem examination on the victims of your own neglect at an earlier stage.

The abolition of gravel from the hatching-boxes has certainly tended to diminish the fish-culturist's opportunities of making himself familiar with the diseases to which the trout is subject. Prevention is better than cure, and it is now well understood that the grand preventive of disease among the trout is cleanliness. Under the old system in which gravel and artificial 'hides' played such a conspicuous part, cleanliness, as it is now understood, was an impossibility ; but with the hatching-boxes of modern design there is no excuse for contaminated water.

Just as improved sanitary conditions have banished many forms of human disease from among us and relegated them to the pages of history, so the modern arrangements of a well-managed fishery have entirely got rid of many of the epidemic scourges which used to harass the breeder and decimate his fry. But lest the beginner, lulled by immunity into a sense of false security, should grow inattentive to the care of his little charges, let me mention some of the maladies to which trout life is exposed.

Livingston Stone mentions no less than twenty-three different troubles, and I will give you his grim catalogue, which, long as it is, does not profess to include all the ills that trout-flesh is heir to.

1. Fungus on the egg.
2. Partial suffocation.
3. Strangulation of the embryo in hatching.
4. Seth Green's dropsy, or blue swelling.
5. Deformity at birth.
6. Fungus on the surface of the body.
7. Constitutional weakness.
8. Emaciation.
9. Starvation.
10. Ulcers on the head.
11. Animal parasites.
12. Fin disease.
13. Black ophthalmia.
14. Irritation of the optic nerve.
15. Inflammation of the gills.
16. Black gill fever.
17. Fatty degeneration of the vitals.
18. Spotted rash.
19. Strangulation by food.
20. Cannibalism, nibbling.
21. Overheating.
22. Suffocation.
23. Paralysis.

All these are very undesirable acquaintances, and I am happy to say that, hitherto, very few of them have obtruded on me their unwelcome presence at the Weston Fishery.

Stone says, ' I suppose there never was in the world such a thing as a fish doctor ; ' and the curative treatment of trout disease certainly shows now as little prospect as ever of becoming a branch of medical science. Fungus (*Saprolegnia*) has, indeed, proved amenable to medical treatment. The patient was immersed in salt water until it turned on its back, this briny bath being repeated at intervals 'as before,' until the invalid was convalescent. The aid of surgery, too, is said to have been invoked with success when the life of an alevin, suffering from Seth Green's dropsy, was saved by a surgical operation which involved 'tapping the sac and letting out the dropsical matter.' But it is a general rule that, when once a young trout has shown signs of being attacked by disease of any kind, the fate of that trout is sealed.

The encouraging fact, however, about these maladies is that the most serious of them are capable of being absolutely prevented; but you must not expect to enjoy this complete immunity unless you have a suitable water supply, and modern apparatus, and unless the eggs and fish receive, in every stage, the constant care and attention which is due to them.

At the end of eight weeks from beginning to feed, the fry should have made considerable growth, and should be sufficiently strong to shift for themselves in protected water.

They can now be turned out for stocking lakes,

rivers, and streams with much less fear of the very heavy loss from starvation which occurs among the very young fry when turned into the wide, wide world straight from the snug nursery of the hatching-box.

Of course, in stocking a stream or lake with fry, even though they are three or four months old, and are forward in growth, a large percentage of them may generally be expected to disappear from other causes than starvation. Many will be devoured by larger trout. Perch and eels, too, will claim a good many victims. Great numbers, too, may disappear in a flood. It follows that, although the first cost of fry is less than that of older fish, to stock with fry is not usually the cheaper plan in the end. On the other hand, there are exceptional circumstances in which stocking even with very young fry may undoubtedly be tried with advantage and economy.

In carefully fenced water containing no fish whatever, a good return may sometimes be obtained from stocking with fry. But this is a branch of my subject in which it is not possible to lay down general rules of universal application. So much depends on the circumstances of each case, that you will generally be well advised in taking the opinion of an experienced fish-culturist before incurring any considerable outlay in stocking with fry.

In cases where it is thought desirable to adopt
this method of stocking, great care should be exercised
in selecting the spot at which to liberate the fry. A
good plan is to take them to a small shallow tributary
of the stream or piece of water which is to be stocked,
and if you can find a place on this rivulet where
watercress grows on either side, turn the fry out here ;
not all at once, but a few at a time.

Assuming, however, that you have decided to
grow your fish to yearlings, you must now transplant
the fry from the boxes into suitable rearing-ponds
prepared beforehand to receive them. I propose to
offer a few hints as to these ponds, before passing on
to the treatment of the fish which have to spend in
them so important a part of their existence.

PONDS

Brook-water, as I have said, is more suitable than
spring-water for rearing purposes. By spring-water is
meant water taken straight from the spring: it is
then cold and barren and subject to little variation of
temperature, but let it run only for a few yards in its
natural course and a new character has been acquired.
The water has taken in oxygen from the air, its temperature has changed, and it is now charged with
animalculæ : it has, in fact, become brook-water.

The very properties which made brook-water unsuitable for egg hatching, are those which will do yeoman service when your hatching-trays have been laid aside for the year, and your attention is turned to rearing. Water, therefore, which contains an abundance of natural food, such as water-shrimps, larvæ of insects, water-snails, and other more minute forms of animal life, is the best for rearing-ponds.

The quantity of natural food which a stream or piece of water will be found to contain depends largely on the nature of the soil which forms the bed, and on the character of the water for supporting the vegetable life on which the insects themselves thrive and multiply.

In the case of some large natural ponds in the south of England, the conditions are so favourable to the growth of water-insects, crustacea, &c., that it has been found possible to rear annually a limited number of trout on a commercial scale without the help of artificial feeding. But even in such exceptional cases the number of trout reared is very small when compared with the number which the same pond could yield if the supply of natural food were supplemented by artificial feeding.

It is sometimes a good plan, however, if your supply of natural food is deficient, to increase it by setting apart ditches and accessory ponds connected with the rearing-ponds for the cultivation of water-insects, &c. But even then you will probably not be able to dispense with the aid of artificial feeding if you would secure results which would be deemed satisfactory by a fish-culturist.

In considering whether the water from which you propose to supply your rearing-ponds is suitable, do not forget the warning which I gave you at the outset against any risk of uncontrollable floods, over-heating in the summer, pollution, or, worst of all, failure or interruption of the supply.

'The best engineer is he who can turn Nature to the best account.' So said Francis Francis when comparing two widely different systems of trout-rearing ponds, each apparently well adapted to the peculiar conditions of its own surroundings.

The remark is singularly appropriate. In turning to account old pieces of water so as to make them productive, in selecting the situation of new ponds, and in considering how best to construct them, whether for breeding or for sport, for profit or for pleasure, there is scope for the exercise of much care, forethought, and engineering skill. There will be natural advantages of situation to be utilised, local difficulties to be overcome, economy to be studied, and regard to be paid to the amenities of the place.

So far from being a disfigurement to your grounds, a set of trout-ponds well constructed and properly attended to will be found to constitute one of the ornamental features of the locality, affording at the same time a subject of continued interest and pleasure to yourself and to your friends.

Part of an ancient moat which used to encircle the old Hall at Weston still exists Until recently the moat, like so many ponds in this country, was choked with mud, weeds, and decaying leaves, the abode of a few coarse fish, and a great number of eels. As to the cooking qualities of the old inhabitants of this water, the less said the better. The typical pond-carp was once defined as 'typhoid reduced to a scaly creature.'[1] All this is altered now. The removal of the unwholesome accumulation of ages disclosed a beautiful, gravelly bed. But a more startling revelation was in store, when, excavating for the foundations of the brickwork of a new outlet, the skeleton form, as it were, of a sluice of an ancient design was discovered. This curious relic of a bygone age seemed to speak reproachfully of the time when private ponds were, like those attached to

[1] *The Practical Management of Fisheries*, by Francis Francis.

the monasteries, properly kept up and cultivated and were made to produce well-tasting fish during the seasons of fasting.

The Weston moat has once again become 'a thing of beauty,' and, as I hope, 'a joy for ever.' At any rate it produces annually a rich harvest of fat two-year-old trout, whose gambols on a summer evening are certainly more enlivening than the torpid move-ments of a carp, which though extolled by Isaac Walton as 'the queen of rivers, a stately, a good, and a very subtil fish,'[1] has so far fallen from her high estate that, according to an eminent writer of modern times, she may often 'be seen lying among the weeds at the surface of the water at which she lazily sucks, making a very distinct noise which has been likened to that produced by a pig.'[2]

I have said that a set of trout-ponds should be one of the attractive features of your place, but I will give you a word of warning, and being on dangerous ground, I will screen myself behind high authority.

'If the reader must have a pair of swans to look pretty, let him get a skilful taxidermist to stuff him a pair as life-like as possible ; he may even, to render the illusion more real, put some clockwork

[1] *The Complete Angler.*
[2] *The Fishes of Great Britain and Ireland,* by Francis Day.

in them to make them nod their heads every five minutes ; let him then anchor them far out in the middle of his pond, and for all picturesque purposes his end will be answered. Knowing the mischief they do, I detest swans, and I wish every swan was, like his sable brother, a *rara avis in terris*, or rather *aquis*. One had better throw open his pond or river to all the poachers in the district than indulge in a taste for swans.' [1]

But to return to the subject of construction of ponds.

'It is not, as a rule, a good plan by damming up a stream to make ponds that may vary in volume.' [2] So says Seth Green, and the rule has been found very useful in warning beginners, inexperienced in the disastrous results of flooding. The problem, however, of dealing with surplus flood water is one with which most fish-culturists are now familiar.

On a small rivulet running through your property, and not liable to floods, a series of ponds can be formed by damming up the stream at intervals, and this is frequently the best plan to adopt in such cases ; and even with streams which are subject to floods effectual provision can, in most instances, be made

[1] *Fish Culture*, by Francis Francis.
[2] *Home Fishing and Home Waters.*

for carrying off all surplus water, and the leaves and rubbish brought down by it, before they reach the intake of your ponds. This is done by a side channel or by-wash of ample capacity.

But perhaps the most common plan is to form a chain of ponds at the side of a brook from which the water supply is drawn, leaving the surplus water to pass off by the natural channel of the stream. In some situations it is a matter of great difficulty to get the water up, or the ponds down, to a suitable level for securing the necessary fall for a series of ponds.

In such cases the services of a practical engineer may have to be called in ; for it is impossible to devise a general scheme which would be well adapted to the peculiarities of each and every case.

The proof of the pudding is in the eating ; and, perfect as a system may appear on paper, it will be of no use unless it effectually secures your object, which is to ensure that the supply of water passing through the breeding ponds shall be constant and sufficient through all the vicissitudes of a fickle climate.

In determining the number of fish a rearing-pond should hold, the size of the pond is of less importance than the minimum quantity of water which passes through it. But the temperature requires to be carefully observed, especially in the summer ; and should the water show a tendency to become heated, the supply must be increased or you will run a serious risk of loss.

To have complete control over the supply, and to be able to create an artificial 'spate' by sending down a flush of additional brook-water into your ponds whenever you think it desirable, is a very great advantage.

The use of a Hornsby-Akroyd oil pumping-engine at the Weston Fishery, for the purpose of increasing at will the natural supply of brook-water during the hottest days of the summer, has been attended with great success.

In respect to size, your ponds should be small for rearing purposes. 'When you mean business,' says Livingston Stone, 'build your ponds small; never let a trout escape to any place where you cannot get at it, observe it, and capture it at a moment's notice.'

The shape of your rearing-ponds will depend upon circumstances; generally it is convenient to have them oblong. If the water supply is not very plentiful or cool the ponds should be deep and narrow, exposing a smaller surface to the rays of the sun.

The quantity of water entering your ponds may be regulated by a hatch at the intake. In front of this is a sloping screen of perforated zinc designed to

SOME TWO YEAR OLD PONDS

prevent the inlet from becoming clogged by leaves, weeds, and other floating matter collected and drawn against it. The outlet of a yearling pond should be so placed as to ensure free circulation for the current through the entire length and breadth of the pond.

Two-year-olds require less current, but their ponds should be wider and deeper, although not necessarily longer than the yearling ponds. It must not, however, be forgotten that a wider pond exposes a larger surface to the sun's rays, and if there is danger of over-heating, shade must be provided.

A convenient arrangement for a series of ponds is where they are separated by perpendicular dams or sluices formed by water-boards which slide into the grooves of a frame firmly fixed in a heading of brick-work, as shown in the illustration (fig. 16). By means of these boards the height of water in each pond can be easily regulated.

To the topmost board of each sluice is attached a shoot of charred wood. To head up the water and increase the capacity of a pond you simply pull out the top board of the outlet sluice, and before replacing it insert an additional water-board of the required width.

The outfall of each pond forms, as will be seen, a waterfall at the inlet of the pond immediately below it. The greater the fall from one pond to the other

P

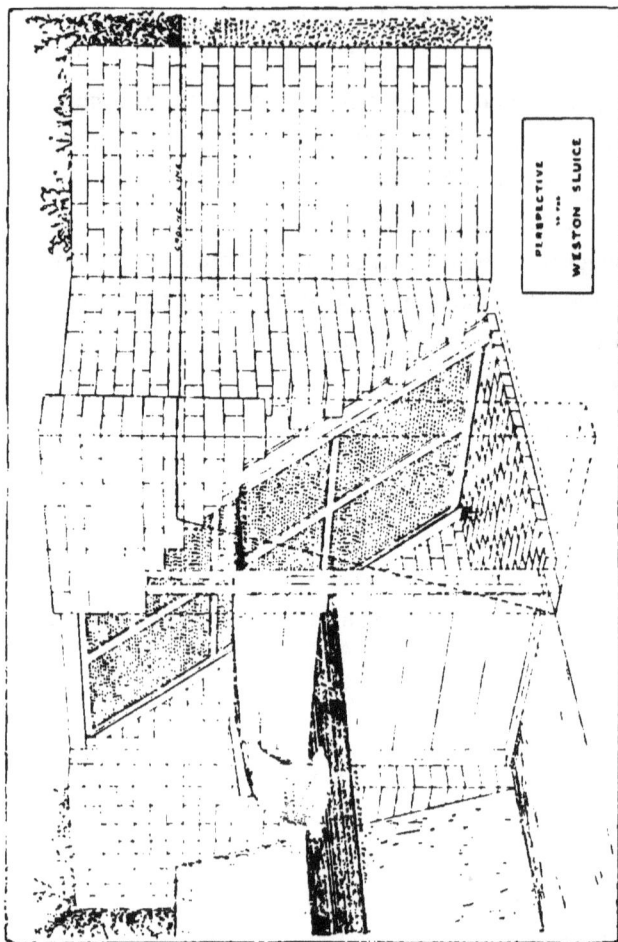

PERSPECTIVE
of the
WESTON SLUICE

FIG. 16

the better for the fish. Much care should be taken in adjusting the shoots. When they are working properly you will see an effervescence of bubbles rising to the surface far down the pond in a line with the current. This shows that plenty of air is being carried into the water.

To prevent the fish from escaping over the outlet a screen is inserted in front of each dam. These screens are made of perforated zinc stretched on strong frames of oak which reach quite to the bottom of the brick floor of the outlet. They slide in grooves and are made to fit closely, but not so tightly that they cannot be drawn out when necessary.

As with the screens in the hatching-boxes, so with the screens for the ponds, it is advisable to have them sloping with their lower end up-stream. In this position they are more easily cleaned, and there is no longer the old danger of any weak fish being forcibly sucked against a perpendicular outlet screen.

In the perspective (fig. 16) a portion of the brick-work is supposed to have been removed, in order to expose the screen and to show the mode in which the water-boards are inserted, and the levels of the water.

The importance of having the screens 'fish-tight' will be most fully realised when, through a small hole or an ill fitting frame, many hundreds of your trout

P 2

have escaped, or when perhaps one or two fish of larger size have been let loose amongst their small cousins, and have pursued, possibly for some weeks, their wicked cannibalistic career unnoticed and unchecked.

The python who swallowed his companion in the 'Zoo' died of indigestion. Like the villain in the story, he came to a bad end. It is not so with the cannibal trout. His digestion being sound and his conscience tough, he seems to grow more fat and prosperous with each murder he commits. His sleek condition certainly does not suggest the downward path, but rather the idea of living happily ever afterwards, to enjoy the fruits of his hardened villainy. This is unsatisfactory. But there is a moral. Don't let it occur again ! Be very careful in future to have tight screens.

After the water has passed through several rearing-ponds it is a very good plan to give it a breathing space, by allowing it to ripple for some distance in a shallow raceway, ' with many a silver waterbreak above the golden gravel.' The raceway should be shaded, and should contain no fish. By this means the water is aërated, cooled, and purified before entering the next series of ponds.

Shade is good for trout in ail stages. In the early

days of fish culture shelter was provided by fixed
hiding-places at the bottom of the ponds. It was
considered 'more natural' that the fish might, as
Buckland put it, 'creep under them and lie all together like rats under a barn floor.' The use of these
fixed 'hides' has, however, been condemned by fish-
culturists as a delusion and a snare. It is undesirable,
as I have pointed out, that any of the fish should be
able to conceal themselves permanently from view.
Your object is to make them grow rapidly ; and to do
this they must be educated to rise readily to their food
and not to be always dozing when dinner is served.
But a more serious count in the indictment against
'hides' is that they harbour dirt ; and this is a very
grievous offence, where no effort is being spared to
promote cleanliness. The necessary shade can, however, be secured in several other ways, as, for example,
by natural position, by trees or shrubs, by high
banks, by deep ponds, by movable covers, by water-
plants, by marginal vegetation. The only objection
to marginal trees and shrubs is that the falling leaves
give extra trouble in the autumn in keeping the
screens clear.

The ponds should be so constructed that you can
run them quite dry whenever required. It is only by
doing this that you can be absolutely certain that one

or two fish are not lurking behind ready to devour
any smaller fish put in. At the end of the season it
is well before re-stocking to clear out all mud and
sediment and to allow the ponds to remain empty
for some time, the longer the better. The exposure
to the air and sun has a very wholesome effect.

It is an excellent plan to introduce into your
ponds suitable water-vegetation by planting it at
intervals on the bottom. 'Water-plants,' says Stone,
'consume carbon and return oxygen. Trout consume
oxygen and return carbon. By putting plants and
fish together, therefore, we avail ourselves of one of
Nature's great universal agencies, in balancing vital
forces against each other, and maintaining the equilibrium on which the continuance of organic life
depends.'

There are, however, many unsuitable kinds of
water-plants. What you want is vegetation which
will afford plenty of shade, and which will form a
suitable 'cover' and pasture for insects, snails, &c.
'Ill weeds grow apace;' and it must not be a plant
which will give trouble by developing a rank growth
and choking the pond.

I know of nothing better adapted to the purpose
than the water star-wort (*Callitriche verna*). Three or
four clumps of it flourish in each of the rearing-ponds

DRAGGING TWO YEAR-OLD POND

at Weston. Set on the bottom with a large piece of
clay in the centre, a tuft of the plant will grow up
compactly ; and the sight of the fish clustering together
in the shadow thrown by this light green pillar of in-
terwoven foliage, as it sways gently with the current,
is one to gladden the heart of the breeder. Water
lake-wort and water lobelia are also excellent plants
for the ponds. So is common watercress, especially
for the shallower parts.

In addition to water-vegetation natural food may
be introduced with advantage, especially into new
ponds. Almost all the forms of pond life are very
prolific, and will be found to increase with great
rapidity. The common water-snail (*Limnaea Peregra*)
is a great favourite with trout. But it is *Gammarus
Pulex*, the freshwater shrimp, which takes the palm.
' I would leave no stone unturned to introduce them
in large numbers,' said Francis Francis. 'When I
have found it to exist in any quantity I have invariably observed the trout are of fine size, and in
unusually good condition.'

This was written thirty-four years ago. At the
end of July 1897, I received a letter from a friend in
which, referring to a consignment of yearling rainbow
trout sent to him only six or seven months previously,
he says : ''They have thriven magnificently, and

some of them *now* are not far short of ¾ lb. in weight.
It is an interesting fact that those which inhabit the
bed of the old brook, the waters of which have been
raised by dams and which swarm with freshwater
shrimps, are almost as pink as salmon, while those
which have been placed in a stream which runs
through a newly made bed, and which are more
dependent than the others on the fly, are nearly pale
yellow.   This seems to bear out what Francis Francis
used to say as to pink flesh being due to freshwater
shrimp.'

My correspondent's observation also furnishes a
striking commentary on a passage in Walton's 'Complete Angler,' 'If I catch a trout in one meadow, he
shall be white and faint, and as certainly if I catch a
trout in the next meadow, he shall be strong and red,
and lusty, and much better meat.'

One more word before I take leave of *Gammarus
Pulex*. He follows a humble but useful calling,
which has endeared him to the fish-culturist.   He is
an indefatigable scavenger.

Trout-water can also be usefully stocked with the
larvæ of various insects which are more familiar to us
in the winged stage of their existence.   At the head
of the list of these larvæ you will not be surprised to
find that of an old familiar friend, the Mayfly.

GROWING THE YEARLINGS AND
TWO-YEAR-OLDS

Shortly before removing the fry from the rearing-boxes, the ponds which are to receive the little tenants should be headed up, but only to the height of about a foot and a half. I do not think it is a good plan to head up the ponds long beforehand, because time would then be given for the weed to grow up. Useful as it will prove at a later stage for harbouring snails, shrimps, &c., when the fry have grown big enough to eat them, water-vegetation is not desirable just at this critical juncture. The fry are not yet big enough to 'assimilate' the snails and shrimps, and there is great danger lest the timid little creatures should, when first liberated, conceal themselves in the weeds with the inevitable result — starvation.

When everything is ready for their reception, the fry must be removed from the rearing-boxes, and put in possession of their more spacious and dignified abode. It has often been observed that young fish are keenly sensitive to the influences of wind and temperature, and certainly it is not a good plan to turn out fry when there is a bitterly cold wind blowing. Especially is this so if the situation of your pond is such that the surface of the water is exposed to the wind prevailing at the time.

The removal is accomplished with the help of syphon and fry-carrier as on the previous occasion when in early babyhood the fish emerged from the hatching-boxes. The whole operation of transplanting the fry and turning them out into the yearling ponds is one which requires care and judgment. The fish, having received in the boxes a sound education in the art of feeding, are not indeed likely to hide or sulk in large numbers on the bottom and there die of starvation, as they frequently do if turned out into ponds at an earlier stage; but the little fellows will certainly at first feel strange in their new home and not much inclined to feed.

Now is the critical period when especial care and constant attention are required. The fish must on no account be subjected to any sudden disturbance, and for this reason it is well not to allow visitors to inspect the ponds just at this juncture. With careful management, however, the fry will not be long in recalling the lesson learnt in the rearing-boxes, and you will enjoy the satisfaction of seeing them darting up at the food with all their accustomed eagerness.

The same diet and system of feeding is continued as in the rearing-boxes, but a long-handled feeding-spoon is necessary (fig. 17). This is a simple contrivance made of a strip of perforated zinc of the

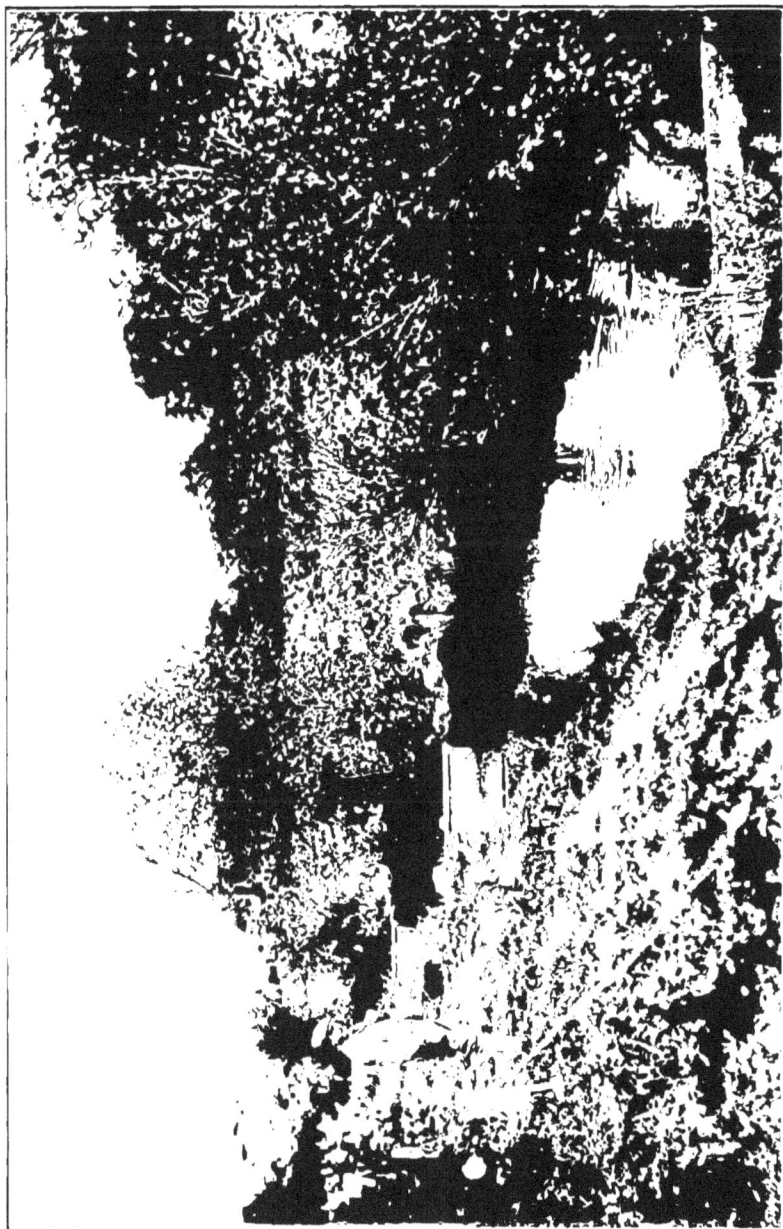

FEEDING YEARLINGS

same mesh as the sieve through which the food has
been pressed, wrapped round a circular base of
wood. The lower end of the handle is flattened and
the cylinder is attached to it by a screw, a small
wooden wedge being inserted to tilt the cylinder to
such an angle as will make it stand vertical when
lowered into the water. A piece of thick galvanised
wire firmly attached to the handle and passing round
the cylinder, as shown in the illustration, serves to
keep feeder and handle from parting company in
spite of continual shaking.

Fig. 17.

With careful management the little fish will soon
come to recognise their old friend the feeding-spoon
and to yield to its fascination. A few magic passes
of this enchanter's wand, and you will seem to have
the power of a wizard to 'summon spirits from the
vasty deep.' But this happy state of things is not
brought about except by regular and systematic
attention. If the fish cannot receive this, then the
result of your attempts at fish culture will be dis-
appointing.

Pigs are said to fatten better when there are several in the same sty, because rivalry, overcoming repletion, prompts each pig to fresh efforts at the trough if only he can prevent his brothers from getting what would have remained uneaten had he supped alone, and in this way the appetites of all are stimulated.

The behaviour of young trout in a pond is very similar. An experienced attendant takes full advantage of this jealous and quarrelsome disposition. He allows only a small quantity of food to be washed out at each dip of the long-handled feeding-spoon. A scramble ensues, loiterers are attracted to the spot, and the habit of collecting to feed at the surface is thus confirmed. To strengthen this habit with his tiny pupils is one of the most important duties of the breeder ; for unless the practice is learnt quickly many of the fry will receive a check in their growth which will mean undersized yearlings in the autumn.

If ever you have been an eye-witness to the tireless efforts of a skilled attendant, coaxing and wheedling the little fish to feed soon after they have been turned out into the ponds, you will not be disposed to begrudge the enviable results which are the outcome of so much unremitting attention.

In order that all the fish in a pond may get their

share of the food the long-handled feeder must be worked, at intervals of a few feet, along the whole length of the pond at each feeding. Should you neglect this essential point, the more robust fish, generally to be found congregating round the inlet, will get the lion's share, and their weaker brethren, as in the outside world, will go to the wall. The former, taking advantage of their increased bulk and strength, will assume bullying and tyrannical ways, making it a difficult matter for their less fortunate companions to get a due share of nourishment. The result of this tyranny, if not counteracted, is that eventually the disparity of size becomes such that cannibalistic proclivities assert themselves and you will have the mortification of seeing one of your most forward and promising pupils swimming about with the tail of his murdered little sister sticking out of his mouth.

You will now understand that if you wish to think kindly of the dispositions of your small fry you must not expect too much of them. 'Drink fair,' was as we know the unwritten law of the gin bottle among the friends of Mrs. Gamp. But among trout there is no pretence of wishing to 'eat fair.' Beauty of form and elegance of movement are theirs; but not the higher graces of character and conduct. It is their rule that the weakest go to the wall, for they are firm believers of the Darwinian doctrine of the survival of the fittest.

Take pains, then, in feeding to distribute the food, and to see that any weaklings near the outlet-screen receive special attention. And this brings me to one of the most serious objections to automatic feeding machines. They go on discharging food regularly enough ; but all of it in one place. The stronger fish therefore collect round the feeder and prevent the others from getting a proper share.

Six or seven times a day is not too often to feed the fish in the ponds at first. Gradually, however, as the fish grow larger, their meals are less frequent, and the zinc of the sieve and also of the feeder are once more changed, to suit bigger mouths and stronger digestions.

A change of diet will now soon be necessary. Food of a more solid and sustaining character is beginning to be required. Sheep's plucks and horse-flesh are probably the best foods available for your purpose. Having first been well boiled, the meat must be finely minced.

The illustration (fig. 18) is of an excellent American mincing-machine, made by the ' Enterprise' Company of Philadelphia. It would be difficult to find a machine better adapted to your purpose

although there are several other suitable mincers in
the market. By changing the perforated steel plate
of the machine the meat is finely
or coarsely cut according to the
size of the fish to be fed. The
perforations of the fine plate which
is used at first are only $\frac{3}{16}$ of an
inch in diameter.

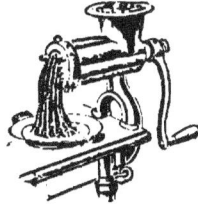

FIG. 18.

The special processes belonging
to the preparation of the food are such that there
should be a separate shed or outbuilding devoted
exclusively to the service of the commissariat depart-
ment. In vain will you have incurred the trouble
and expense of constructing rearing-ponds if you have
not also made proper arrangements for provisioning
the fish to be put into them. You will be like the archi-
tect of a dwelling-house, who had exhausted so much
skill and attention on the dining-room that he forgot
to provide for the kitchen, the scullery, and the larder.

The minced meat having been allowed to cool,
and having been separated by rubbing in the hand, is
taken in a small pail, and scattered broadcast evenly
in the ponds. From this time the use of the long-
handled feeder is discontinued. A complete and
permanent alteration from one diet to another should
not be made suddenly. The prepared liver food

is not therefore at first wholly stopped, but is thrown to the fish with the minced meat, the proportion of liver being gradually reduced until it finally disappears altogether from the menu.

' Little and often ' is the motto of the successful fish-culturist in relation to feeding in the earlier stages. But as the fish grow larger the labour of feeding becomes less continuous, and by the end of August two meals a day will generally be found sufficient. A greater quantity of food, however, is now consumed per day in two meals than previously disappeared when the fish were fed more frequently. This is as it should be.

The task of keeping the screens in front of the sluices clear of leaves and rubbish forms a most important part of the work connected with a well-managed fishery. They must be frequently swept with a long-handled brush (fig. 19). The slope of the screens greatly facilitates this operation. You must be careful, however, to avoid injuring any of the young fish by catching them between the brush and the screen.

During the spring and summer months trout cannot as a rule be safely transported. On a cool day in September I have sometimes ' travelled ' yearlings successfully, but the risk is great, and it is not before

you are well into November that, generally, the tem
perature will be low enough to make it safe to move
trout to any considerable distance.

As soon as the fish can be safely travelled, they
are for the first time dignified by the name of 'yearlings,' although in reality they are then for the most
part only about ten months old, and by the name of
yearlings they continue to be known until the approach of next spring closes their travelling season,
and gives the signal for their removal to the larger
ponds where they will be treated as growing two-year-
olds.

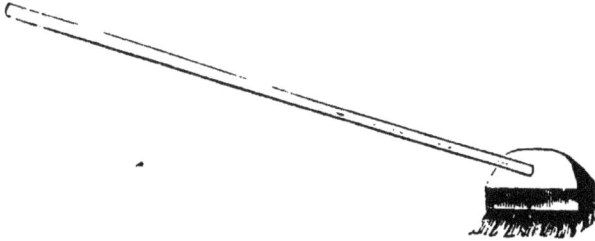

Fig. 1)

The departure of the yearlings enables you to run
the yearling ponds quite dry, ready to be cleaned and
'aired' before receiving the next stocking of feeding
fry from the rearing-boxes.

The management of the growing two-year-olds is
merely a continuation of the treatment which the fish

Q

were receiving when they left the yearling ponds.
The plate of the mincing-machine will, however, require to be replaced by one with perforations $\frac{3}{8}$ths
of an inch in diameter instead of $\frac{3}{16}$ths, and the
amount of food thrown into the pond at each visit
must be gradually increased to keep pace with the
feeding capacity of the fish.

It is only experience which will teach you exactly
how much food to give to a certain number of fish
of a certain age. The quantity varies greatly. The
temperature of the water, the state of the atmosphere,
the season of the year, the amount of natural food—
these and other considerations have to be taken
into account.

The different breeds of fish also feed differently.
The rainbow trout (*Irideus*), although a late spawner,
has a more voracious appetite and a quicker growth
than either the English brook trout (*Fario*) or the
Loch Leven trout (*Levenensis*); consequently a pond
full of the rainbow should be fed more generously
than the same number of their English or Scotch
contemporaries in an adjoining pond.

'I think it safe to say,' says Livingston Stone,
'that under favourable circumstances large trout of
any age will eat one-fiftieth of their weight per diem
in the summer, that one per cent. of their weight a

day will keep them in good condition through the
year, and that they would do very well on half that
allowance. And,' he adds, 'I have also observed
that with two-year-olds five pounds of meat food is
an equivalent for one pound of trout growth.'

A summer's evening is perhaps the best time for
the visitor who wishes to 'see the animals fed.' In
the winter a good opportunity occurs at noon on a
bright sunny day. Almost before the food has left
your hand the surface of the water will begin to
'boil,' and the food has hardly touched the water
before the commotion is such as to suggest the idea
of a seething cauldron. Not an atom of the food
will be allowed to reach the bottom, and many of the
fish shooting up like rockets, will hurl themselves
clean out of the water in their eagerness to avoid
being late for dinner.

Especially is this so with those champion surface-
feeders, the rainbow trout, which, by their manner of
feeding alone, you would soon learn to distinguish
with your eyes shut. The commotion made by the
simultaneous lashing of the surface-water marks
Irideus as the possessor of that vigour and pluck
which is beginning to make him such a favourite with
the fly-fisher.

Q 2

TRAVELLING

The conveyance of live trout even to considerable distances is not now the risky and difficult matter which it was once thought to be.   Indeed it is nowadays quite the exception for any loss to occur in transit even on a very long train journey, so well are the conditions of successful travelling understood.

Before being despatched the fish must undergo a course of preparation for the journey.   Accordingly they are captured by means of a net made of coarsely woven peach-netting.   This fabric is admirably adapted to the present purpose ; it is soft and pliant and contains no harsh chafing knots, nor are there any entangling meshes in which the gill covers of the fish can sustain injury.

The shape of the net and the size of it will depend on the shape and size of the pond.   In the case of some very small ponds a dip net may be found sufficient.   This net resembles a very large two-handed landing-net with a rectangular fixed iron frame. The illustration facing p. 230 will assist you to imagine an implement which, although it would hardly form an acceptable addition to the outfit of the gentle angler, is no less indispensable to the breeder than is his crook to the shepherd.   But a drag-net or seine is generally

DRAGGING YEARLING POND WITH POLE NET

necessary for the present purpose. It should be made of ample length so as to be very 'baggy' when in use; leads should be fixed along the bottom, and corks along the top. For the yearling ponds, which are narrower than those containing the older fish, a drag-net of special design called a pole-net is found very convenient. Instead of being drawn by ropes this net is stretched on two side-poles of considerable length. By this means the bottom corners of the net are kept down, as it follows the poles which are pushed along the sides of the pond. If the water is frozen the net must be drawn under the ice, a passage being cut for the side poles or ropes along both sides of the pond.

In removing the fish from the net, great care is required to prevent them from being injured. They should on no account be touched with the hand, but should be emptied out into baths, and carried at once to a preparing tank or box through which a good supply of water is kept running. Here they remain in durance vile, without any food for two or three days in the case of yearlings, and for a longer term in the case of two-year-olds. In either case the length of the preparation required depends, to some extent, on the distance to be travelled, and the temperature of the water.

After these days of total abstinence and confinement the fish will be in prime condition for travelling, and you will thus avoid the necessity of changing the water during the journey, a change which has so often proved fatal.

Yearlings and two-year-olds, unlike the fry, travel best in frosty weather. When on the point of departure the fish are transferred to the large specially made zinc carriers which appear in the illustration. Fish-carriers are constructed according to various patterns, many of them very ingenious, and some of them very fanciful in design. The principle of them all is to make the shaking and vibration of the vehicle do the work of aëration. The water is allowed to splash about in such a way that enough fresh oxygen is taken in from the air to compensate for the oxygen taken out of the water by the fish's gills.

Ice has been used for travelling fish in warm weather, but I do not recommend it. The objection is that when, at the end of their journey, the fish are suddenly turned out into water of higher temperature, there is some danger of inflammation of the gills resulting.

A very serviceable fish-carrier, for want of anything better, is to be found in one of those large,

REMOVING YEARLINGS FROM PREPARING TANK FOR TRAVELLING

glass wicker-clad bottles which are used for conveying commercial sulphuric acid, and are known as carboys. The bottle should not be more than three parts full, otherwise the process of automatic aëration, so to speak, would cease, and the fish would be suffocated. The risks involved in allowing the water to be changed *en route* are great. The character of the new water is often unsuitable, and a sudden alteration in temperature is calculated to bring on gill fever. In the case of a large consignment going a considerable distance, where there are several changes of trains, a competent attendant should accompany the fish for the whole, or at least a part, of the way to make assurance doubly sure. In the early days of fish culture the tobaccoless attendant was fondly supposed to aërate the water with great vigour and little intermission by plying a pair of bellows during the whole of a long railway journey.

No sooner are the carriers filled than they are hoisted on a cart standing in readiness to convey them to the railway station. The cart is timed to reach the station a few minutes before the departure of the train by which the fish are to travel. Careful arrangements have of course been made with the consignee by which the fish are to be met at the end of their railway journey, and conveyed from the station without delay to the water which they are intended to stock, and there liberated.

Experience shows that if the precautions I have mentioned are observed, the fish are none the worse for their experiences in overland transport.

The larger the fish the greater of course is the quantity of water required for travelling them. The expense, therefore, of transporting two-year-olds is proportionately heavier than in the case of yearlings, and the risk is more appreciable.

The danger of overcrowding in the cans is greater for rainbow trout than for *Fario*, although, when at liberty, rainbow have 'broken the record' for thriving in water which is subject to considerable variation of temperature.

One of the perils of travelling is, lest the carriers, on a change of trains, should by accident be left standing still exposed to the heat of the sun for some time ; and precautions must be taken to prevent such an accident as this from happening.

As I have said, any great alteration of temperature is injurious if it is very sudden. Accordingly, before liberating the fish, it is necessary gradually to pour from the carrier the water in which they have travelled, substituting water from the stream or lake which you are stocking. This is especially desirable where there is a difference of several degrees between the temperature of the water in the carrier and that of the water into which the fish are being turned out.

Never discharge the trout into the new water from a height of several feet. If in a thoughtless moment you should do this, many of them will come to the surface on their backs and expire gasping from the effects of the sudden shock.

After such a misfortune it is to be hoped that the consignee may not meet the hard fate of the father of the unhappy Dina in the old song. He, as you will remember, came himself to a bad end after having been haunted by the reproachful spectres of his defunct victims, saying, 'We shouldn't have died if it hadn't been for you.'

CHAPTER III

STOCKING

STOCKING is a branch of my subject on which alone a treatise could be written.

The fish-culturist is constantly being asked to advise as to the best and cheapest means of stocking a stream, a lake, or a pond. General rules are of very little use here. A great variety of local conditions have to be taken into consideration, and it is hardly possible to give reliable advice and useful recommendations except on full information as to the facts and circumstances of each particular case. Indeed, it is often found that a personal inspection of the water by an experienced pisciculturist is necessary in order to secure a thoroughly satisfactory return for a given outlay.

Where immediate sport is required two-year-olds are recommended, and where it is found impossible to get rid of the coarse fish, or where older trout already exist, two-year-olds are often the least expensive mode of stocking, although the first cost is much heavier than that of yearlings.

Two-year-old fish are better able to hold their own against pike, which frequently play havoc amongst yearlings.

The trout should be monarch of all he surveys. All coarse fish should therefore, wherever possible, be made to give place to him. Although roach and dace do not cause direct loss except to the ova, their room is decidedly more valuable than their company in trout water. They consume very large quantities of natural food which the trout would otherwise get, and generally I have no hesitation in advising the removal of all coarse fish even where their presence does not seem to do harm.

Two-year-olds require deeper water than yearlings, and if put in very shallow parts of the stream they tend to work downwards. As a general rule, however, and especially where the water is free from pike, yearlings easily take the palm for stocking purposes. Well-grown yearlings can, for the most part, be trusted to hold their own against perch; they are excellent travellers, and they readily become acclimatised to new water.

The cost of purchasing or breeding them is small when compared with the cost of older fish. The expense of transporting yearlings by rail too is comparatively small, as they require less water than two-year-olds, and there is seldom any necessity for a consignment to be accompanied by an attendant.

Yearlings should, if well grown, range from three to six inches as early as November. Many of them should be more than six inches, but very few indeed should be under three inches. This, at any rate, is my experience in Norfolk with the eggs of selected breeders, although I am aware that the growth of yearlings reared in higher latitudes is not so rapid.

As to the number of fish of a certain age which can be advantageously used annually or otherwise to stock a given lake or stretch of water, it is not possible to lay down any hard-and-fast rule. Questions as to acreage, depth, character and volume of water, temperature, natural food supply, presence of coarse fish, and a host of other questions have to be considered, and the existing conditions studied.

Should the number of trout introduced be greater than the natural food will support, they will not thrive without the help of artificial feeding, any more than cattle on an overstocked pasture can be expected to fatten on the grass alone.

It is often found that a very heavy stocking of trout in an old piece of water of limited extent will thrive wonderfully for the first year without any extra feeding, but that later on the fish begin to fall off in condition. This is probably due to the partial exhaustion of the supply of natural food. The balance of life has been disturbed and must be restored. The natural food supply can probably be replenished by cultivating the water and introducing snails, shrimps, &c., and in the meantime the help of artificial feeding should be called in.

A new pond or lake recently constructed, and containing no water-vegetation and little natural food, must not be expected at once to support a large quantity of good-sized fish in prime condition, unless the cultivation of the water is carefully attended to, and in some cases artificial feeding also provided.

It is surprising, however, to find how large a number of trout can be made to thrive and fatten for the rod or table in a small space if due attention is paid to the cultivation of the water and development of its natural resources.

The three principal breeds of fresh water trout raised by the British fish-culturist are the English brook trout (*Salmo fario*), the Loch Leven trout (*Salmo Levenensis*), and last, but not least, the rainbow trout (*Salmo irideus*).

Fario, the English brown trout, is, as he deserves

to be, the most popular variety for general stocking purposes in this country. For streams and rivers his precedence, though sometimes questioned by the pretensions of a new favourite, has been steadily maintained, and as the ‘lusty trout’ of our brooks he seems to be in little danger of ceasing to hold against all comers the place of honour in the estimation of our poets, our sportsmen, and our cooks.

Salmo Levenensis is the lake trout of Scotland and the north of England. He is recommended for stocking lakes, ponds and reservoirs, but it is a mistake to suppose that he will not also thrive in rivers, especially where there is considerable depth of water. It is however charged against him that when turned into unfenced water he shows a disposition to roam, especially down the stream, and that he is dissatisfied to remain where he has been turned in if there is any chance for him to get away.

I am not myself inclined to lay much stress on this supposed characteristic of the Loch Leven trout ; my own view is that he does not differ very widely in habit from other varieties, but his high reputation has undoubtedly been won in lakes rather than in streams. Give him wide scope and plenty of natural food, and his high qualities will show themselves to the best advantage, as they do in the native loch from which he takes his name.

In eulogising the trout of Loch Leven much has been said and written about the flavour, the firmness, and the pinkness of his flesh. I hope I shall not be accused of wounding 'national susceptibilities' when I say that the question is sometimes asked how far these qualities are retained in places where natural food, and particularly the water-shrimp, is not abundant. It would however be a dangerous heresy to dispute the claim of *Levenensis* to share with *Fario* the honour of having earned for his species the proud title of 'the venison of the waters.'

That beautiful variety, the rainbow trout, came to us originally from America. His brilliant hues have fairly won for him the name he bears, while, in point of shape, I might almost say that he satisfies the trout-breeder's ideal of perfection.

If *Levenensis* has been known in some waters to develop a tendency to grow long and slim; the rainbow's habit of growth is invariably short and thick.

A splendid surface feeder, he will readily accommodate himself to new water and will, under favourable conditions, grow with extraordinary rapidity and to a great size. It is not an uncommon thing for rainbow yearlings to be caught three-quarters of a pound in weight before the end of the ensuing summer, viz. when only 14 or 15 months old.

As a sporting fish I do not think *Irideus* could be surpassed. His boldness in rising to the fly and his fighting mettle must be seen and felt to be fully realised and appreciated. The limits of my own personal experience do not at present enable me to speak with certainty as to his suitability for streams ; but for stocking lakes, ponds, and reservoirs I can recommend him with great confidence.

I have never known a case where rainbow trout have on my recommendation been introduced without giving unqualified satisfaction, and in several cases I have myself had an opportunity of testing with rod and fly their high sporting quality.

As remarked in the last chapter, rainbow (and particularly the two-year-olds) are somewhat tender in travelling, and they are found to require rather more 'elbow room' in the carriers than *Fario* ; but on the other hand the possible range of temperature for their habitat is wider, and they have been known to grow and flourish in waters which have not proved cool enough for common trout.

Admirers of *Irideus* have not been able to claim for him total exemption from the common frailty of the Salmonidæ family—the propensity towards cannibalism ; but it is urged, and as I think justly, that he is 'less of a cannibal' than *Fario*. This, however, does not mark *Fario* as the possessor of a larger share of original sin. The fact is that rainbow has the smaller mouth.

As regards flavour and firmness of flesh, it is giving *Irideus* very high praise to say that under favourable conditions he is not outdone even by his British cousins.

CONCLUSION

In conclusion let me give one or two words of advice to the beginner.

Fish culture is a practical science. Diligent study of the best books will yield much useful information, but it is by personal experience alone that you will be enabled to apply successfully the knowledge acquired by reading and to achieve good results.

Be content therefore with small and inexpensive beginnings, and be prepared to learn not only from your successes, but from your failures. Fish culture is no exception to the rule that the more trouble you take the larger will be your measure of success ; but the converse does not hold good : if you take little trouble in trout breeding, you will meet, not with little success, but more probably with complete failure.

R

Negligence and inattention frequently result in wholesale loss—loss which can hardly be retrieved or compensated for by any amount of extra trouble taken afterwards, for the season is gone.

'Almost any one,' says Stone, 'in a favourable locality can raise trout enough to feel rewarded for his pains.' But do not imagine that success can be secured without a sustained effort.

To obtain good results trouble must be taken, and there is no doubt that the best mental equipment for a fish-culturist is the capacity for taking pains.

COOKERY OF THE TROUT

BY

ALEXANDER INNES SHAND

'THE trout is a fish highly valued both in this and foreign nations; he may be justly said, as the old poet says of wine, and as we English say of venison, to be a generous fish; a fish that is so like the buck that he also has his seasons; for it is observed that he comes and goes out of season with the stag and buck. . . . He may justly contend with all sea-fish for precedency and daintiness of taste, and that being in right season, the most dainty palates have allowed precedency to him.' So speaks old Izaac Walton on the fish of his predilection, and, like the orator who addressed the Bristol mob after Mr. Burke, we 'say ditto' to the venerated apostle of angling. Izaac knew what he was writing about, and never writes of anything as to which he knows nothing. We dare not imagine the sublimity of eloquence to which he might have soared, had he played, and landed, and cooked the salmon. But as the salmon disputes precedency with the turbot, so the trout when most

delicate and at his best, rivals or surpasses the red
mullet. If he is handicapped by the absence of 'a
trail,' that is his misfortune, and nature is to blame.
The difference for the epicurean between salmon and
trout is this, that as to the former you may safely
generalise; if not obviously out of season and condition, he is always good, if not super-excellent. The
salmon, like the sea-trout, not only is brought up in
the sea, but he goes back to the brine to recuperate
after each summer of inland degeneration. The trout,
on the contrary, except that unexceptionable *Salmo
trutta*, passes a lifetime in the same locality; he is of
various species, and depends altogether for his flavour
on his particular food and the circumstances of his
surroundings. All the waters in the northern hemi-
sphere teem with trout of all sorts and sizes, from the
bleak little tarn lying in the dark solitudes of Highland
hills, to the broad estuary where the ocean surges are
breaking in foam over the shelving sand bars. There
are the trout of the tarn that multiply mysteriously
on the most anti-Malthusian principles, for to all
appearance there is next to nothing to eat. No Mayfly ever disports itself on those sub-Arctic waters,
cheerless as Athabasca or the Great Slave Lake. Not
many of the amphibious birds fly thither to take toll
of the spawn; and the numbers are only slightly kept

in check by the occasional swoop of the osprey. As may be supposed, these troutlets are chiefly skin and bone, but they are sweet and savoury so far as they go. It is pitiful to see how greedily they will jump at the flies trailing in the wake of the artificial otter. Yet we believe the starvelings would make a delicate dish were they dressed by a clever artist *à la blanchaille*. They are not much bigger than whitebait, early in the season, nor would they come much short of the inimitable sardines on the Mediterranean coasts if fried in batter and served with frizzled parsley. There is a marked improvement in the trout of the moorland burn which has its chilly sources in the tarns. The otter, sated with salmon and sea-trout, the most fastidious of four-footed gourmets, deems it worth his while to include them in his nocturnal prowlings, and the dreamy heron, ever intent upon business, is often to be seen on the rocky shelf overhanging some limpid pool. Between those Highland rills and the chalk streams of the southland counties, we come upon trout of many sorts and conditions. They may be generally estimated from the gastronomic point of view by shape and colouring, for size is no criterion. We have seen a two-pounder taken out of a drain on a rushy common, and we should have been sorry to condemn our worst enemy

to eat him.   Local palates profess patriotic admira-
tion for the bloated monsters, familiar acquaintances
of habitual anglers, who have eluded for years the
parochial poachers.   When one of them is landed at
last and borne off to the village alehouse, he is made
the object of a triumph, and the excuse for a carouse.
But what can you expect of a fish which, like some
sedentary men of letters, has never taken a yard of
exercise, or swam a stroke when he could help it?
He has been battening all his indolent days on
newts and tadpoles and fouler garbage ; and lurking,
when dyspeptical and off the feed, in the slime under
the tangled roots of some alder copse suggestive of
a malarious mangrove swamp in the sweltering delta
of the oil rivers.   Test him by his shape, and he
has run to head and stomach, narrow-shouldered,
and pot-bellied, black and leprous—almost loathsome.
How different from his lively congeners who run
out the line with a salmon-rush, and have fought to
the death before they were landing-netted !   The
choicest trout have the graceful proportions of the
aquatic athlete, and though the tints and speckling
may be varied, they are invariably rich and lustrous.
For trout change their tints like the chameleon,
though less suddenly, and faithfully reflect the colours
and characteristics of their feeding grounds.   They

are in perfection in the rapid rivers flowing through
gravelly pastoral uplands or springing in clean-
stemmed pine forests. Such are the Tweed and
its tributaries, the Spey, or the Aberdeenshire Dee.
The rushes and broken waterfalls find them constant
exercise, while in backwaters and in the quiet swirls
behind rocks and stones, there are resting-places
where they can digest and fatten in tranquillity. From
the nature of the country, there are fine shades of
difference, but all easily distinguishable by the
expert. There is no exception to be taken to a
trout from Tay, hooked anywhere between Scone
and Taymouth. Yet he is distinctly inferior in
flavour—though possibly fuller in flesh—to his
cousins that have been caught in Tweed or Teviot,
because the Tay, after its filtering in the loch of the
name, drains the rich arable farms of the Carse of
Gowrie. For that, as we have remarked, there is an
evident reason. But in other cases, the difference is
as clear, and yet there is no satisfactory explanation.
Why should the heavy brown trout in the lakes of
West Sutherland be unmistakably better than those
in the lakes to the east, while those bred in Loch
Shin are inferior to all the others? Why should the
trout in three adjacent lochs in Arisaig, all stocked
some five and thirty years ago from Loch Morar,

have changed since then in quality as well as in colour? Yet though we believe before all in the trout from the Scottish and Irish lochs and rivers, we are bound to say that some of the Southern streams are scarcely, if at all, behind them.   The Kennet is good, where the fish are more gracefully shaped and more brilliantly speckled as you approach the upland sources ; the Test is still better, where in water crystal clear you lure the wary quarry with gossamer gut, and stalk him from afar on bended knee, care-ful of the faintest fall of your shadow.   There the transparent water runs over glistening chalk, and you might almost read the superscription on a three-penny-bit at two fathoms depth.   But the trout of some Hertfordshire streams perhaps take pre-cedence ; from the Colne, where Sir Humphry Davy's pupil caught the phenomenal fish which his host sent as a present to His Royal Highness the Duke of Sussex ; and, above all, from the oddly named and less known Mimram, which flows through Lord Cowper's preserves of Panshanger.

Probably among all the inland trout of these islands, from the bronze-backed denizens of the northern Highland lochs to the bright-sided beauties of the South, radiant with glowing crimson or golden spots, those of Lochleven are at once the richest and

most delicate. It is agreed now that the genealogy
should be traced to the sea-trout. The bright red
colour is appetizing in itself, and they have invariably
fetched the highest prices in the market. *Laudatores
temporis acti*, we are inclined to think the finest were
even finer than they are now, before the Loch was
professionally *exploité* by the angling clubs, and the
voracious pike had been more systematically netted.
For there is no question that a fair admixture of
pike improves the size, the shape, and the quality of
the trout who have survived to share their feeding
grounds. As to rivers, we have praised the trout
of the Tweed, and it is impossible to read the
description of the memorable dinner at the Cleikum
Inn, in the introduction to Meg Dod's Cookery Book,
without the revival of many a pleasant recollection of
the water springing to the lips. Mrs. Johnston was
a popular novelist in her day, but we venture to
say she never wrote anything more sympathetically
spirited than her story of the institution of the
Cleikum Club, an inimitable gastronomic sequel to
'St. Ronan's Well.' Touchwood was entertaining
the famous Dr. Redgill, a divine as devoted to
culinary science and practice as Dr. Opimian of
'Gryll Grange.' The nabob, sitting at the lavish table
he had spread in Innerleithen, had quoted the ancient

rhyme about the fasting of the jovial monks of Melrose. He asked the Doctor if he remembered the old stave.

'The Doctor remembered no such thing. His attention was given to more substantial doctrine. "Sir, I should not be surprised if they possessed the original receipt—a local one too, I am told—for dressing the red trout in this hereditary house of entertainment?"'

' Never doubt it, man—claret, butter, and spiceries. Zounds, I have eat of it till—it makes my mouth water yet. As the French adage goes, "Give your trout a bottle of good wine, a lump of butter and spice, and tell me how you like him." Excellent trout in this very house—got in the Friar's Cast, man, the best reach of the river. Let them alone for that. Those jolly monks knew something of the mystery.'

It says much for the full flavour and rich essences of the red trout, that a refined epicure like the nabob deemed them improved by such *prononcé* treatment as red wine and spice. Yet we are told that 'the stewed trout had ceased to be a luxury to him,' and we cannot wonder that with long familiarity they had palled on his palate. Good trout, like salmon, are less likely to pall when they are dressed with severe and Arcadian simplicity. For permanent

appreciation of a plain dish, one should stick to
boiling or frying, and that was the opinion of Sir
Humphry Davy, who was as much the philosopher
with the fish-kettles as over his chemical furnaces.
When he had just given elaborate instructions as
to the *al fresco* crimping of the salmon, he says
decidedly as to a noble trout, in answer to a question
from the inquisitive Physicus, who was always eager
for instruction, 'We will have him fried.' And at the
dinner afterwards, when he righteously forbade Harvey
Sauce for the salmon, he only admitted for the trout
a little vinegar and mustard—in fact, the elements
of an *à la Tartare* without the onions. And the
Ettrick Shepherd, who kept open house above lone
St. Mary's Loch, between Ettrick and Yarrow, in a
streamland then almost fabulously *fischreich*, pro-
tested at a banquet in Ambrose's blue parlour 'that
he devoured dizzens and dizzens every week in
the family, maistly done in the pan, with plenty o'
fresh butter and roun' meal.' The poet of the forest
had the root of the matter in him. So we have had
them at many an improvised dinner or supper, in
many a Highland or Irish hostelry, having generally
taken the precaution of being our own trout purveyor.
The Irish bacon might be a trifle rancid, and Irish
potatoes are too often impostures, superbly mealy on

the outside, but sadly waxy within. The Highland
hill mutton was not unfrequently unmistakable
'braxy'—from some wether that had come to an
untimely end, to be snatched at the eleventh hour from
the beaks of eagles and ravens. But in the deserts
the dish of trout was always a sure stand-by, even
when the muirfowl were out of season, and the
poachers had had ill-luck with the red-deer. We have
fallen back, in a caprice, with no small satisfaction
upon cookery even more simple and primitive than
that. It was a blazing day on the moors; the scent
was bad ; the dogs were dead beat ; there was nothing
to be done till the sun sloped down and the birds
were stirring again for the afternoon feeding. The
cool, shrunken brook, trickling down between the
gravel banks scooped out by the floods, looked
irresistibly tempting. So the shooting-jacket was cast
off, the shirt-sleeves rolled up to the armpits, and we
went guddling and groping beneath banks and stones.
The troutlets were tossed walloping into the heather
almost by handfuls. The heather-roots were dry as
tinder, and there was soon what would have been a
cheery blaze, had it not been dimmed by the sun-glare.
It burned down as fast as it blazed up, and on a couch
of soaked bracken, above the smouldering embers,
the trout they had been split and gutted—were laid

down to broil. Some connoisseurs used to dress them
with a sprinkling of gunpowder, but that was matter
of taste.

That was merely the freak of a sated Apicius or
Cambacérès, for there were cold game and other
dainties in the panniers. But how often have anglers
camping out in the wilds relied on their rods for what
the worthy Mrs. Bertram of Ellangowan would have
called the best part of their dinner?   In Scandinavia,
the Canadian Dominion or British Columbia, the
pièce de résistance should have been bear-meat, reindeer-
haunch or saddle of bighorn, as the case might be. The
hunter might often come back to camp empty-handed,
and the place of the joint was a melancholy blank ; but
the angler had filled possible-sack and pockets, and
trout, fried or boiled, figured conspicuously in the
menu. Those roving sportsmen should have said
grace with heartfelt gratitude over the fish which agree-
ably varied their diet, yet after all, if reduced to short
commons, they had only to turn and go back.   But
the trout have played no insignificant part in the
daring exploration of the North American continent.
They saved many a valuable life, with others which,
perhaps, were scarcely worth the saving.   Lewis and
Clarke, Fremont and other official adventurers were
followed through the passes of the Rockies and the

unknown wilderness beyond by the reckless pioneers
of the rival fur companies. Many of these men, at
the peril of their lives and at the cost of inconceivable
privations, had acquired invaluable information which
must have perished with them. Suffering from cold
and nakedness, with famine hourly staring them in
the face, the tribes of Indians, starving themselves,
refused to part with either horseflesh, dried roots, or
dog-meat. There were sinister stories as to missing
men who had been sacrificed by lot for the sake of
their companions. What is certain and well authenti-
cated is, that when they had taken some brawling
river for their guide, except for the occasional trap-
ping of a beaver or a muskrat, their lives were saved
solely by trout. To the trout the topographical ex-
plorers of the Great West are even more indebted
than to the black buck or the buffalo.

The trout used always to be endeared to the
gastronomic tourist as the *spécialité* of many Continental inns and hotels. We said we had frequently
been indebted to our rod for a supper in Scottish or
Irish inns. On the Continent we have found it much
the reverse; we have often come back to some famous
fishing quarter, with a basketful of beauties which
had given short but splendid play in the streams
of Tyrol or Salzkammergut; they were taken over

with contemptuous civility, sent down to the kitchen, and we have more than a suspicion that they never reappeared. The fact is that in these lands of piscatorial abundance, the hosts swear by their *viviers*, and we rather think they are right. The 'vivier' is an artificial tank or a great wooden chest moored in the stream by heavy stones. The sides of the chest are pierced with holes, and the water flows freely through. It is regularly supplied by anglers or poachers in the pay of the establishment. The fresh-caught trout are brought in from the neighbouring waters in barrels with a convex side fitting to the shoulders. In these *viviers* the prisoners are fed or gorged with chopped livers and other kitchen refuse. It is the stuffing of an Alsatian goose without the pain or the liver disease, for the fish in tip-top condition seem always voraciously on the feed. When they are wanted for table, the cook takes a survey and makes his selection. It is said that the practice was bequeathed by the monks, and if so, the gastronomist is indebted to them for the legacy. Pleasant recollections we have of meals and merry picnics at the hotel at the Eaux Chaudes, where, looking down on the pool in the Gave de Gabas from the balcony of the *salle-à-manger*, you saw your dinner, as the Scotch say,-- in superb vitality. Yet those anchored *viviers* had their mishaps like our

s

lightships, and were sometimes the cause of bitter disappointment. We remember one morning at a venerable *gasthaus* in Tyrol, .the host coming in clutching at his cropped hair, and seemingly on the brink of a suicide like that of M. Vatel. The river had come down in spate during the night, and the fish-chest had gone floating down with it. And by the way, when we tried comfort on that occasion, by reminding him of the noble trout we had brought home the day before, he made a hasty exit, speechless and embarrassed. If we are not greatly mistaken those beautiful trout of ours had been chopped small and cast out as food for the living.

In Switzerland, before the railways had brought excursionists in battalions, from Basle to Lugano the trout was everywhere as much the *plat de maison* as the *chamois* in the Bernese Oberland or the Savoy Alps. Trout were expected at all repasts as much as a matter of course as fresh herrings in an Hebridean steamer in the tourist season. *Apropos* to that, Dumas, in his delicious 'Impressions de Voyage en Suisse,' tells a comically pathetic tale in his own most characteristic manner. He had got to Bex, in the Valais, and ordered trout for breakfast. He was told the manner of catching them was peculiar, so perhaps he might like to see it. Always on the alert for a sensation, or a novelty, he jumped at the offer. He followed his sulky guide in the dark by the dim light of a lantern. The man stripped to the waist ; plunged into the glacier-water with lantern and sharpened hook, and as the trout circled up to the light, like moths fluttering to a candle, the deadly stroke was dealt.   The '*pauvre diable*,' resigned to his fate, had no illusion as to his end : as the last hunter from the inn had broken his neck when after chamois, so he, like his fishing predecessor, was sure to die of an inflammation.   Naturally, Maurice held the trout in no favour.   'It's not only you that are fond of them' he said bitterly, when Dumas entreated him to go back to his blankets.   'I cannot conceive how it is, but all the travellers like the trout, a nasty brute skinful of bones.'   Another illustrious French tourist who followed Dumas, agreed with Maurice, though through sheer cross-grainedness. There is no more *piquant* scene in the memorable Swiss tour of About's Captain Bitterlin than where the Captain calls the unlucky waiter on the Rigi across the coals, when his perversities chanced to culminate and the everlasting trout were the pretext.

'Est-ce que vous vous moquez de moi ? You gave me trout yesterday morning at the hotel in Basle, and I ate them.   Last night at Olten, still more trout, and I said nothing.   This morning at Lucerne

s 2

a third dish of trout ; I do remark upon it, and here
you are again, all the same this evening ! You have
sworn then to turn me into a trout, and that is the
respect you pay to my observations.'

Nowadays, neither a M. Bitterlin nor anybody
else need complain of being overdone with trout in
Switzerland.   There may still occasionally be fair
draughts of fishes, but what are they among so
many?   And we venture to assert that, even in days
gone by, very few tourists have been fortunate
enough to taste the trout of Lake Leman. We have
spent weeks, or rather many months, on the Lake,
and have only seen them once or twice at a *table
d'hôte*. Yet the *Salmo Lemanus* is no whit inferior
to the freshest of salmon trout sheathed in the sea-
lice, and well deserves to be memorialised, as we
may presume he has been, by that admirable dressing, *à la Génervoise*. And before taking leave of the
Continent we must pay a grateful tribute to the
attractions of the trout of the Wolfbrunnen at
Heidelberg—the sylvan fountain which is said to
take its name from the sorceress who fell a victim to
the were-wolf.   One remembers the student suppers
made uproarious with jest and song, where the fishes
from the five ponds, after the walk through the woods
from the Schloss, were the motive of the boisterously

jovial meeting. Nor ought we to forget Vaucluse, which we venture to say has been visited by many pilgrims, and by not a few illustrious masters of the culinary art, rather for the eels, the exquisite trout and the *écrevisses*, than for the sake of the sonnets and the sentimental associations.

Urban Dubois is fanatical in his enthusiasm for the principle of the *viviers*. Virtually he lays down the law that a trout is not worth eating unless 'cooked alive.' By which we do not understand that it should be treated with such brutal inhumanity as the eels in Italy or the shell-fish in England, but only that it should leap straight from the water to the fire. Though, after all, a dip in a boiling kettle is as easy a death as any. But we should like to ask M. Dubois if he ever extended his researches to Scotland or Norway. As we remarked when writing of the salmon, the sea-trout runs him hard, and there are epicures who give the preference to the lesser fish. It is obviously impossible to improve on perfection, and we maintain that nothing with fins can be much more perfect than the *Salmo trutta*, fresh run from sea or fjord, though he may necessarily have been landed some few hours before. In that case, away with the condiments and sauces, and serve with nothing but the water impregnated with brine!

When the sea-trout are running inland, so full are they of vigour and vitality, that they come complacently to the very door of the Gastronomer located in the far interior. Drawing less water than the bulky salmon, they rush straight up river and streamlet without a check, unless the ascent is barred by some impracticable cataract. They make a journey of a hundred miles or more in less than a couple of days, and you may take them still shining in their silvery sheathing among the small brown trout in the moorland burn. Of course when you land the leviathan in these embarrassing circumstances he should be inspected before being cooked. He may have been cramped and confined for weeks in the tiny pool, after the subsidence of some violent spate, like the hippopotamuses, crocodiles, and minor eccentricities which Baker describes as observing armed neutrality in some deep reach of the Atbara, when after the descent of the Abyssinian floods half the abounding river has been drawn up by the blazing sun and the other half swallowed in the thirsty sands. For the sprightly sea-trout, in these untoward and insanitary surroundings, falls off fast in flesh and condition, and becomes the most poverty-stricken of finny creatures.

As to the *Salmo ferox*, he gives noble sport, but

compared to other trout of the lakes, he is worth but
little for the table. The dull red flesh of the cannibal
par excellence with the sinister name, is inclined to
choke you off, should you see him split in the kitchen.
But even with him there are differences, and the
most edible are to be found in Loch Awe and the
lakes in Western Sutherland, with the single exception of Loch Ericht—the loftiest lake of any of
its size in the British Islands—which may be taken to
be the natural home of the *Ferox*, and where he is of
unrivalled strength, though of moderate proportions.
The gillaroo, another variety of the lake-trout, to be
caught in North Irish and West Highland waters, is
better flavoured, though of less formidable shapes.
Like the bull-trout—the *Salmo erix*—of the borders,
it is broad across the shoulders and thick in proportion to its length. And by way of postscript we may
mention there is one notable peculiarity in the Nor-
wegian trout as compared to the Scottish. The latter
when in season seem to be unaffected by temperature.
With the Norwegians, on the contrary, when the
temperature gets much above 56°, they lose flavour
and condition with startling rapidity.

For the processes of cookery, on the principle
of *seniores priores*, we give precedence to Izaac
Walton's old-fashioned recipe.

'Take your trout, wash and dry him with a clean napkin; then open him, and having taken out his guts and all the blood, wipe him very clean within, but wash him not, and give him three scotches with a knife to the bone, on one side only. After which take a clean kettle, and put in as much hard stale beer—but it must not be dead—as will cover the fish you intend to boil, then throw into the liquor a good quantity of salt, the rind of a lemon, a handful of sliced horse-radish root, with a handsome little faggot of rosemary, thyme and winter savoury. Then set your kettle upon a quick fire of wood, and let your liquor boil up to the height before you put in your fish; and then if there be many, put them in, one by one, that they may not so cool the liquor as to make it full; and whilst your fish is boiling, beat up the butter for your sauce with a ladleful or two of the liquor it is boiling in; and being boiled enough, immediately pour the liquor from the fish, and being laid in a dish, pour your butter upon it, and strewing it plentifully over with shaved horse-radish, and a little pounded ginger, garnish the sides of your dish, and the fish itself, with a sliced lemon or two, and serve up.'

We don't dare to criticise the theology of a St. Paul or a St. Augustine, yet even after the nabob of

St. Ronan's recipe for the red trout of Tweed, Izaac seems to go somewhat far in the spiceries. But he knew the trout of the Lea in the seventeenth century better than we do, and his prescription has a romantic, old-world savour, like his sheets fragrant with lavender. Some respectable authorities say the sea-trout should never be boiled, in which we differ : *toujours* the broil must become tiresome at last. But when the trout are boiled, simplicity should be studied. 'Put the fish into nearly boiling water, with a tablespoonful of salt. Boil up quickly, then simmer till done. For a pound, allow twelve minutes ; for two pounds, fifteen minutes ; for three pounds, twenty minutes. If the trout be unexceptionable serve with melted butter.' If, as Mrs. Poyser has it, you must seek the flavour in the cruets, you may break out in a broad variety of sauces. But undoubtedly broiling and frying are the preferable methods of treatment. To fry small trout —clean, wash and dry— roll in flour, and fry in butter or clarified dripping to a delicate brown, and do not overdo. Use a heated cloth to absorb the grease, and serve with frizzled parsley and slices of lemon With larger fish, rub with flour or oatmeal, and fry in a deep pan, immersing them in butter, clarified dripping, or lard, heated so that the fish shall neither be scorched nor stewed. Or, if they are to be broiled,

split, wipe very dry, dust, and use a clear, moderate
fire. As for baking, we cannot conscientiously recommend it, but here is the method : ' Place the
trout in a deep pan, scatter pieces of butter in it.
Season without and in the inside with allspice, mace
and salt. Baste, from time to time, with what collects
in the basting pan.' If you do bake your fish, we
are inclined to think it is better served cold, when the
fancy may have free range as to condiments, and it
may be sent up with a seasoning of Chili vinegar and
bay leaves.

Here is another and a somewhat simpler recipe
for stewing than Mr. Touchwood's : ' Clean the fish
well. If large, cut them in fillets. Rub inside with
salt and spices ; lay them in a stewpan and cover with
good stock, with two onions, some black peppercorns,
cloves and a bit of mace. When the fish have stewed
for a few minutes, add two glasses of white wine, a
little lemon and cayenne, and perhaps an anchovy.
Add a teaspoonful of mushroom ketchup.' Stewing
is all very well for eels ; but in our judgment it is
scarcely orthodox for delicate trout. We should opine
that the gourmet who orders a trout to be stewed
in spices, condemns it as undeserving more respectful
treatment. Yet we need hardly say we hesitate to
dispute the authority of the monks of Melrose, who

must have been connoisseurs on all cookery on their *jours*. However *maigres*, the trout dressed *au bleu* is something of a compromise. 'Clean the trout, put them in a saucepan in the pint of boiling vinegar which turns them blue. Add as much Chablis or Pouilly with water till the trout are covered. Put in a few bay leaves, salt, pepper, an onion with some cloves, the peel of a lemon, a little celery and parsley, and boil for a quarter of an hour. Serve with oil and vinegar.'

Small and somewhat flavourless trout nevertheless make an excellent souchet, and a souchet, after all, must depend on the appendages for its savour. 'Boil a few parsley roots in a pint and a half of water for five minutes ; add a teaspoonful of grated horse-radish and a teaspoonful of salt. Put in five or six small trout, boil, skim, simmer for twelve minutes. Serve in the water with the parsley, in a deep dish, and send up the thin brown bread and butter, *de rigueur.*' Analogous to the trout *au bleu* is the dressing *à la Génevoise*, doubtless a delicate compliment to those trout of Geneva which we praised. 'Clean, but do not scale. Put a little *court bouillon* in a stewpan with parsley roots, cloves, parsley, bay leaves, onions and a carrot. Stew for an hour, strain the liquor over the trout in a small fish pan with a glass of

Madeira. When boiled, drain and remove the scales;
then put it in the pan with a little of the liquor.
Make a thickening and add veal gravy or red wine;
season with mushrooms, parsley and green onions.
Stew till smooth. Strain the sauce over the dished
fish, with squeeze of lemon, essence of anchovy and
some mace.'

To make a superlative *court bouillon*, season a
gallon of water with salt, carrots, parsley, eschalots,
thyme, bay leaves, half an ounce of peppercorns and
a few cloves. Stew and strain, adding two bottles of
red wine and one of white.' The recipe sounds extravagant, but then the *marinade* may be kept and
used repeatedly, with the addition of a little water.
And we must reiterate that a rich dressing of the sort
should only be used with trout that are either inferior
or somewhat stale.

Yet that is not the opinion of M. Urban Dubois,
who as a Court cook is inclined to swear by the
court bouillon. And M. Dubois ranks the sea-trout
and lake-trout scarcely below the princely salmon
he adores, and he gives some elaborate recipes for
ornamental *plats de luxe*, intended for banquets or
ball suppers. This is his recipe for salmon trout
à la Chambord, which used to be a *spécialité*, both at
Philippe's and the Frères Provençaux. 'The trout is

stuffed and trussed, then fixed on the drainer of a
fish-kettle, taking care to place it on its belly; it is
covered with a good *court bouillon* and cooked in it; the
liquor must be poured out cold on the fish; it is afterwards brought to ebullition, but immediately removed
from the fire. A salmon-trout of about ten pounds
must remain an hour and a quarter in its liquor, kept
at the same degree of heat, but without any visible
ebullition. The trout is dished immediately, resting
on its belly; but for more safety it must be kept in
equilibrium by means of fried bread crusts put so as
to keep the fish upright; this bread must be masked
with raw forcemeat and poached in the oven. The
garnish which surrounds the fish is composed, on
one side, of mushroom heads, whole truffles, small
quenelles: on the other it is composed of carp-milts,
crayfish tails, or a whole crayfish with the tails
trimmed, and lastly with quenelles.' All these
garnishes are slightly masked with Genevoise or
matelote sauce; the remainder is served in a sauce-
boat. It may be observed that, except for purposes
of ostentation, the dish would be better were the
trout half the size. The bigger fish might be turned
to good account for the slices *à la Jeanne d'Arc*.

'Six fine slices compose the entrée: they are
pickled cold, grilled, then dished up on a bread or a

rice *croustade,* carved in a low form with six sides.
This *croustade* bears in its centre a low column in
fried bread or rice, at the top of which is fixed a
small vase in vegetables, cut out, garnished with a
few small quenelles, dished *en rosace,* having a small
truffle in its centre.'

M. Dubois describes with pride and gusto how
he dressed an exceptionally beautiful Neckar trout
'with jelly,' and directs special attention to that
fish's portrait.

'To cook such a trout it must be first cleaned
thoroughly, and filled with an ordinary fish-stuffing ;
the head trussed, and the whole wrapped in a cloth,
then placed on the drainer of a fish-kettle ; it is then
covered with a good cold *court bouillon.* The fish-
kettle is put on the fire till it boils up ; it is then
withdrawn to the side, only allowing it to simmer.
If the trout weigh between 7 and 8 lbs., it must
simmer for about an hour. When done the fish
must be left to get cold in the liquor. The fish-
drainer is then set cross-wise over the kettle, the
trout taken out of the cloth, and kept in this position
for about an hour to drain it well. It must be
sponged, and then directly slid on to an unheated
fish-stand. It is *nappé* with jelly *mi-prise,* and en-
circled with croûtons of jelly.'

A good variation for dressing a fine trout is the *à la Norvège*, one of the culinary triumphs of the late accomplished chef at the Wyndham Club. There the special flavour was given to the sauce which somehow was a subtle reminder of the fjords and the fir woods, by a strong infusion of chopped chives and Tarragon.

Small trout, more or less tasteless, when no better are to be had may be disguised with the luscious stuffing, *à la Condé*. 'They are cut to an equal length, cleared of their bones, filled with a truffle-quenelle forcemeat, and wrapped up separately in a strip of paper. They are cooked in a good stock, but with little liquid, and allowed to cool in their stock. Afterwards they are unfolded, trimmed, warmed at the entrance of the oven, with their residue reduced to a half-set glaze. At the moment of serving the trout are dished in a circular form, in an upright position, on a border of quenelle force-meat, ornamented, poached, and turned out in a dish, in the centre of which is fixed a bread-support, made of dried bread, masked with forcemeat ; on this support a small *croustade* made of bread and paste is solidly fixed. This *croustade* is garnished with a *buisson* of large blanched and trimmed oysters. On each trout is placed a slice of truffle. Serve with

Espagnole sauce, reduced with the residue of the
trout and the trimmings of the truffles.'

Though prime trout should be treated with due
respect and served, as we have said, in the water that
boiled them, sauces come in usefully with the fish
which have little but 'the honour of the name' to
commend them.   With mayonnaise and Tartar one
can never go wrong, the chief difference between
these being that more eggs are mixed with the Tartar,
and there is an addition of chives, green onions, and
shallot with a sprinkling of chopped gherkins.   The
colder it is the better it is.   In Dr. Redgill's *sauce
piquante*, scraped horse-radish, four eschalots, a clove
of garlic, a drachm of mustard, and one of celery-
seed are blended with cayenne and black pepper.
When these are well pounded a half pint of cu-
cumber vinegar, a quarter pint of shallot, and as much
horse-radish, are to be infused, but we need hardly
add that in that case the best thing to be done with
the trout is to throw them away before troubling with
such elaborate dressing.   We may say much the
same of prescriptions for potting, for the best of the
trout part with all that is most characteristic in the
process.

Spottiswoode & Co. Printers, New-street Square, London.

THE BADMINTON LIBRARY
OF SPORTS AND PASTIMES.

Edited by HIS GRACE THE (EIGHTH) DUKE OF BEAUFORT, K.G.
and A. E. T. WATSON.

Crown 8vo. 6s. net each Volume, cloth.

*** *The Volumes are also issued half-bound in leather, with gilt top.*
Price 9s. net each.

ARCHERY.
By C. J. LONGMAN and Col. H. WALROND. With Contributions by
Miss LEGH, Viscount DILLON, &c. With 2 Maps, 23 Plates, and
172 Illustrations in the Text. Crown 8vo. 6s. net.

ATHLETICS.
By MONTAGUE SHEARMAN. With Chapters on Athletics at School by
W. BEACHER THOMAS; Athletic Sports in America by C. H. SHERRILL;
a Contribution on Paper-chasing by W. RYE, and an Introduction by
Sir RICHARD WEBSTER (LORD ALVERSTONE). With 12 Plates and
37 Illustrations in the Text. Crown 8vo. 6s. net.

BIG GAME SHOOTING.
By CLIVE PHILLIPPS-WOLLEY.
Vol. I. AFRICA AND AMERICA. With Contributions by Sir SAMUEL
W. BAKER, W. C. OSWELL, F. C. SELOUS, &c. With 20 Plates and
57 Illustrations in the Text. Crown 8vo. 6s. net.
Vol. II. EUROPE, ASIA, AND THE ARCTIC REGIONS. With Con-
tributions by Lieut.-Colonel R. HEBER PERCY, Major ALGERNON, C. HEBER
PERCY, &c. With 17 Plates and 56 Illustrations in the Text. Cr. 8vo. 6s. net.

BILLIARDS.
By Major W. BROADFOOT, R.E. A. H. BOYD, SYDENHAM DIXON, &c.
With 11 Plates, 19 Illustrations in the Text, and numerous Diagrams.
Crown 8vo. 6s. net.

COURSING AND FALCONRY.
Coursing. By HARDING COX, thoroughly Revised by CHARLES RICHARD-
SON.—*Falconry.* By the Hon. GERALD LASCELLES. With 20 Plates and
55 Illustrations in the Text. Crown 8vo. 6s. net.

CRICKET.
By A. G. STEEL, the Hon. R. H. LYTTELTON, ANDREW LANG, W. G. GRACE,
&c. With 13 Plates and 52 Illustrations in the Text. Crown 8vo. 6s. net.

CYCLING.
By the EARL OF ALBEMARLE and G. LACY HILLIER. With 19 Plates
and 44 Illustrations in the Text. Crown 8vo. 6s. net.

DANCING.
By Mrs. LILLY GROVE, F.R.G.S. Miss MIDDLETON, the Hon. Mrs.
ARMYTAGE, &c. With Musical Examples, and 38 Full-page Plates and
93 Illustrations in the Text. Crown 8vo. 6s. net.

LONGMANS, GREEN, & CO. LONDON, NEW YORK, AND BOMBAY.

THE BADMINTON LIBRARY
OF SPORTS AND PASTIMES.

Edited by HIS GRACE THE (EIGHTH) DUKE OF BEAUFORT, K.G. and
A. E. T. WATSON.

Crown 8vo. 6s. net each Volume, cloth.

. *The Volumes are also issued half-bound in leather, with gilt top. Price 9s. net each.*

DRIVING.
By His Grace the (Eighth) DUKE OF BEAUFORT, K.G., A. E. T. WATSON, the EARL OF ONSLOW, &c. With 12 Plates and 54 Illustrations in the Text. Crown 8vo. 6s. net.

FENCING, BOXING, AND WRESTLING.
By WALTER H. POLLOCK, F. C. GROVE, C. PREVOST, E. B. MITCHELL, and WALTER ARMSTRONG. With 18 Plates and 24 Illustrations in the Text. Crown 8vo. 6s. net.

FISHING.
By H. CHOLMONDELEY-PENNELL.
Vol. I. SALMON AND TROUT. With Contributions by H. R. FRANCIS, Major JOHN P. TRAHERNE, &c. With 9 Plates and numerous Illustrations of Tackle, &c. Crown 8vo. 6s. net.
Vol. II. PIKE AND OTHER COARSE FISH. With Contributions by the (Fourth) MARQUIS OF EXETER, WILLIAM SENIOR, G. CHRISTOPHER DAVIS, &c. With 7 Plates and numerous Illustrations of Tackle, &c. Crown 8vo. 6s. net.

FOOTBALL.
History. By MONTAGUE SHEARMAN.—*The Association Game.* By W. J. OAKLEY and G. O. SMITH.—*The Rugby Union Game.* By FRANK MITCHELL. With other Contributions by R. E. MACNAGHTEN, M. C. KEMP, J. E. VINCENT, WALTER CAMP, and A. SUTHERLAND. With 19 Plates and 35 Illustrations in the Text. Crown 8vo. 6s. net.

GOLF.
By HORACE G. HUTCHINSON. With Contributions by the Right Hon. A. J. BALFOUR, M.P. Sir WALTER SIMPSON, Bart. ANDREW LANG &c. With 32 Plates and 57 Illustrations in the Text. Crown 8vo. 6s. net.

HUNTING.
By His Grace the (Eighth) DUKE OF BEAUFORT, K.G. MOWBRAY MORRIS, the (Eighteenth) EARL OF SUFFOLK AND BERKSHIRE, G. H. LONGMAN, &c. With 5 Plates and 54 Illustrations in the Text. Crown 8vo. 6s. net.

MOUNTAINEERING.
By C. T. DENT, Sir W. M. CONWAY, D. W. FRESHFIELD, &c. With 13 Plates and 95 Illustrations in the Text. Crown 8vo. 6s. net.

POETRY OF SPORT (THE).
Selected by HEDLEY PEEK. With a Chapter on Classical Allusions to Sport by ANDREW LANG, and a Special Preface to the BADMINTON LIBRARY by A. E. T. WATSON. With 32 Plates and 74 Illustrations in the Text. Crown 8vo. 6s. net.

RACING AND STEEPLE-CHASING.
By the (Eighteenth) EARL OF SUFFOLK AND BERKSHIRE, W. G. CRAVEN, the Hon. F. LAWLEY, ARTHUR COVENTRY, and A. E. T. WATSON. With Frontispiece and 56 Illustrations in the Text. Crown 8vo. 6s. net.

LONGMANS, GREEN, & CO. LONDON, NEW YORK, AND BOMBAY.

THE BADMINTON LIBRARY

OF SPORTS AND PASTIMES.

Edited by HIS GRACE THE (EIGHTH) DUKE OF BEAUFORT, K.G. and
A. E. T. WATSON.

Crown 8vo. 6s. net each Volume, cloth.

₊ *The Volumes are also issued half-bound in leather, with gilt top. Price 9s. net each.*

RIDING AND POLO.
By Captain ROBERT WEIR, the (Eighth) DUKE OF BEAUFORT, the EARL OF ONSLOW, &c. With 18 Plates and 41 Illustrations in the Text. Crown 8vo. 6s. net.

ROWING.
By R. P. P. ROWE and C. M. PITMAN. With Chapters on Steering by C. P. SEROCOLD and F. C. BEGG ; Metropolitan Rowing by S. LE BLANC SMITH ; and on PUNTING by P. W. SQUIRE. With 75 Illustrations. Crown 8vo. 6s. net.

SEA FISHING.
By JOHN BICKERDYKE, Sir H. W. GORE-BOOTH, Bart. ALFRED C. HARMSWORTH, and W. SENIOR. With 22 Full-page Plates and 175 Illustrations in the Text. Crown 8vo. 6s. net.

SHOOTING.
Vol. I. FIELD AND COVERT. By Lord WALSINGHAM, Sir RALPH PAYNE-GALLWEY, Bart. the Hon. GERALD LASCELLES, and A. J. STUART-WORTLEY. With 11 Plates and 95 Illustrations in the Text. Crown 8vo. 6s. net.
Vol. II. MOOR AND MARSH. By Lord WALSINGHAM, Sir RALPH PAYNE-GALLWEY, Bart. Lord LOVAT, and Lord CHARLES LENNOX KERR. With 8 Plates and 57 Illustrations in the Text. Crown 8vo. 6s. net.

SKATING, CURLING, TOBOGGANING.
By J. M. HEATHCOTE, C. G. TEBBUTT, T. MAXWELL WITHAM, &c. With 12 Plates and 272 Illustrations in the Text. Crown 8vo. 6s. net.

SWIMMING.
By ARCHIBALD SINCLAIR and WILLIAM HENRY. With 13 Plates and 106 Illustrations in the Text. Crown 8vo. 6s. net.

TENNIS, LAWN TENNIS, RACKETS AND FIVES.
By J. M. and C. G. HEATHCOTE, E. O. PLEYDELL-BOUVERIE, and A. C. AINGER With Contributions by the Hon. A. LYTTELTON, W. C. MARSHALL, Miss L. DOD, &c. With 12 Plates and 67 Illustrations in the Text. Crown 8vo. 6s. net.

YACHTING.
Vol. I. CRUISING, CONSTRUCTION OF YACHTS, YACHT RACING RULES, FITTING-OUT, &c. By Sir EDWARD SULLIVAN, Bart. the (Thirteenth) EARL OF PEMBROKE, LORD BRASSEY, K.C.B. R. T. PRITCHETT, E. F. KNIGHT, &c. With 21 Plates and 93 Illustrations in the Text. Crown 8vo. 6s. net.
Vol. II. YACHT CLUBS, YACHTING IN AMERICA AND THE COLONIES, YACHT RACING, &c. By R. T. PRITCHETT, the MARQUIS OF DUFFERIN AND AVA, K.P. the EARL OF ONSLOW, &c. With 35 Plates and 160 Illustrations in the Text. Crown 8vo. 6s. net.

LONGMANS, GREEN, & CO. LONDON, NEW YORK, AND BOMBAY.

FUR, FEATHER, AND FIN SERIES.

Edited by A. E. T. WATSON.

Crown 8vo. 5s. each Volume, cloth.

* *The Volumes are also issued half-bound in leather, with gilt top.
Price 7s. 6d. net each.*

THE PARTRIDGE.
Natural History. By the Rev. H. A. MACPHERSON.—*Shooting.* By A. J. STUART-WORTLEY.—*Cookery.* By GEORGE SAINTSBURY. With 11 Illustrations and various Diagrams in the Text. Crown 8vo. 5s.

THE GROUSE.
Natural History. By the Rev. H. A. MACPHERSON.—*Shooting.* By A. J. STUART-WORTLEY.—*Cookery.* By GEORGE SAINTSBURY. With 13 Illustrations and various Diagrams in the Text. Crown 8vo. 5s.

THE PHEASANT.
Natural History. By the Rev. H. A. MACPHERSON.—*Shooting.* By A. J. STUART-WORTLEY.—*Cookery.* By ALEXANDER INNES SHAND. With 10 Illustrations and various Diagrams. Crown 8vo. 5s.

THE HARE.
Natural History. By the Rev. H. A. MACPHERSON.—*Shooting.* By the Hon. GERALD LASCELLES.—*Coursing.* By CHARLES RICHARDSON. *Hunting.* By J. S. GIBBONS and G. H. LONGMAN.—*Cookery.* By Col. KENNEY HERBERT. With 9 Illustrations. Crown 8vo. 5s.

THE RABBIT.
By JAMES EDMUND HARTING. With a Chapter on Cookery by ALEXANDER INNES SHAND. With 10 Illustrations. Crown 8vo. 5s.

THE SALMON.
By the Hon A. E. GATHORNE-HARDY. With Chapters on the Law of Salmon Fishing by CLAUD DOUGLAS PENNANT; Cookery, by ALEXANDER INNES SHAND. With 8 Illustrations. Crown 8vo. 5s.

THE TROUT.
By the MARQUESS OF GRANBY. With Chapters on the Breeding of Trout by Col. H. CUSTANCE; and Cookery, by ALEXANDER INNES SHAND. With 12 Illustrations. Crown 8vo. 5s.

PIKE AND PERCH.
By WILLIAM SENIOR ('Redspinner,' Editor of the *Field*). With Chapters by 'JOHN BICKERDYKE' and W. H. POPE. *Cookery.* By ALEXANDER INNES SHAND. With 12 Illustrations. Crown 8vo. 5s.

RED DEER.
Natural History. By the Rev. H. A. MACPHERSON.—*Deer Stalking.* By CAMERON OF LOCHIEL.—*Stag Hunting.* By VISCOUNT EBRINGTON.—*Cookery.* By ALEXANDER INNES SHAND. With 10 Illustrations by J. CHARLTON and A. THORBURN. Crown 8vo. 5s.

LONGMANS, GREEN, & CO. LONDON, NEW YORK, AND BOMBAY.

A Classified Catalogue

OF WORKS IN

GENERAL LITERATURE

PUBLISHED BY

LONGMANS, GREEN, & CO.,

39 PATERNOSTER ROW, LONDON, E.C.

91 AND 93 FIFTH AVENUE, NEW YORK, AND 32 HORNBY ROAD, BOMBAY.

CONTENTS.

History, Politics, Polity, Political Memoirs, etc.

Abbott.—A HISTORY OF GREECE. By EVELYN ABBOTT. M.A., LL.D.
Part I.—From the Earliest Times to the Ionian Revolt. Crown 8vo, 10s. 6d.
Part II. 500-445 B.C. Crown 8vo, 10s. 6d.
Part III. —From the Peace of 445 B.C. to the Fall of the Thirty at Athens in 403 B.C. Crown 8vo, 10s. 6d.

Abbott. — TOMMY CORNSTALK: being Some Account of the Less Notable Features of the South African War from the Point of View of the Australian Ranks. By J. H. M. ABBOTT. Crown 8vo, 5s. net.

Acland and Ransome. A HANDBOOK IN OUTLINE OF THE POLITICAL HISTORY OF ENGLAND TO 1896. Chronologically arranged. By the Right Hon. A. H. DYKE ACLAND and CYRIL RANSOME, M.A. Crown 8vo, 6s.

Allgood. CHINA WAR, 1860: LETTERS AND JOURNALS. By Major-General G. ALLGOOD, C.B., formerly Lieut. G. ALLGOOD, 1st Division China Field Force. With Maps, Plans, and Illustrations. Demy 4to, 12s. 6d. net.

History, Politics, Polity, Political Memoirs, etc.—*continued.*

ANNUAL REGISTER (THE). A Review of Public Events at Home and Abroad, for the year 1901. 8vo, 18s.

Volumes of THE ANNUAL REGISTER for the years 1863-1900 can still be had. 18s. each.

Arnold.—INTRODUCTORY LECTURES ON MODERN HISTORY. By THOMAS ARNOLD, D.D., formerly Head Master of Rugby School. 8vo, 7s. 6d.

Ashbourne.—PITT : SOME CHAPTERS ON HIS LIFE AND TIMES. By the Right Hon. EDWARD GIBSON, Lord ASHBOURNE, Lord Chancellor of Ireland. With 11 Portraits. 8vo, gilt top, 21s.

Ashley.—SURVEYS, HISTORIC AND ECONOMIC : a Volume of Essays. By W. J. ASHLEY, M.A. 8vo, 9s. net.

Bagwell.—IRELAND UNDER THE TUDORS. By RICHARD BAGWELL, LL.D. (3 vols.) Vols. I. and II. From the First Invasion of the Northmen to the year 1578. 8vo, 32s. Vol. III. 1578-1603. 8vo, 18s.

Baillie.—THE ORIENTAL CLUB, AND HANOVER SQUARE. By ALEXANDER F. BAILLIE. With 6 Photogravure Portraits and 8 Full-page Illustrations. Crown 4to, 25s. net.

Besant.—THE HISTORY OF LONDON. By Sir WALTER BESANT. With 74 Illustrations. Crown 8vo, 1s. 9d. Or bound as a School Prize Book, gilt edges, 2s. 6d.

Bright.—A HISTORY OF ENGLAND. By the Rev. J. FRANCK BRIGHT, D.D.

Period I. MEDIEVAL MONARCHY : A.D. 449-1485. Crown 8vo, 4s. 6d.

Period II. PERSONAL MONARCHY. 1485-1688. Crown 8vo, 5s.

Period III. CONSTITUTIONAL MONARCHY. 1689-1837. Crown 8vo, 7s. 6d.

Period IV. THE GROWTH OF DEMOCRACY. 1837-1880. Crown 8vo, 6s.

Bruce.—THE FORWARD POLICY AND ITS RESULTS ; or, Thirty-five Years' Work amongst the Tribes on our North-Western Frontier of India. By RICHARD ISAAC BRUCE, C.I.E. With 28 Illustrations and a Map. 8vo, 15s. net.

Buckle.—HISTORY OF CIVILISATION IN ENGLAND. By HENRY THOMAS BUCKLE. 3 vols. Crown 8vo, 10s. 6d.

Burke.—A HISTORY OF SPAIN from the Earliest Times to the Death of Ferdinand the Catholic. By ULICK RALPH BURKE, M.A. Edited by MARTIN A. S. HUME. With 6 Maps. 2 vols. Crown 8vo, 16s. net.

Caroline, Queen.—CAROLINE THE ILLUSTRIOUS, QUEEN-CONSORT OF GEORGE II. AND SOMETIME QUEEN REGENT : a Study of Her Life and Time. By W. H. WILKINS. M.A., F.S.A., Author of 'The Love of an Uncrowned Queen'. 2 vols. 8vo, 36s.

Chesney.—INDIAN POLITY: a View of the System of Administration in India. By General Sir GEORGE CHESNEY, K.C.B. With Map showing all the Administrative Divisions of British India. 8vo, 21s.

Churchill (WINSTON SPENCER, M.P.). THE RIVER WAR : an Historical Account of the Reconquest of the Soudan. Edited by Colonel F. RHODES, D.S.O. With Photogravure Portrait of Viscount Kitchener of Khartoum, and 22 Maps and Plans. 8vo, 10s. 6d. net.

THE STORY OF THE MALAKAND FIELD FORCE, 1897. With 6 Maps and Plans. Crown 8vo, 3s. 6d.

LONDON TO LADYSMITH VIÂ PRETORIA. Crown 8vo, 6s.

IAN HAMILTON'S MARCH. With Portrait of Major-General Sir Ian Hamilton, and 10 Maps and Plans. Crown 8vo, 6s.

History, Politics, Polity, Political Memoirs, etc.—*continued.*

Corbett (JULIAN S.).

DRAKE AND THE TUDOR NAVY; with a History of the Rise of England as a Maritime Power. With Portraits, Illustrations and Maps. 2 vols. Cr. 8vo, 16s.

THE SUCCESSORS OF DRAKE. With 4 Portraits (2 Photogravures) and 12 Maps and Plans. 8vo, 21s.

Creighton (M., D.D., late Lord Bishop of London).

A HISTORY OF THE PAPACY FROM THE GREAT SCHISM TO THE SACK OF ROME, 1378-1527. 6 vols. Crown 8vo, 5s. net each.

QUEEN ELIZABETH. With Portrait. Crown 8vo, 5s. net.

HISTORICAL ESSAYS AND REVIEWS. Edited by LOUISE CREIGHTON. Crown 8vo, 5s. net.

Dale.—THE PRINCIPLES OF ENGLISH CONSTITUTIONAL HISTORY. By LUCY DALE, late Scholar of Somerville College, Oxford. Crown 8vo, 6s.

De Tocqueville.—DEMOCRACY IN AMERICA. By ALEXIS DE TOCQUEVILLE. Translated by HENRY REEVE, C.B., D.C.L. 2 vols. Crown 8vo, 16s.

Dickinson.—THE DEVELOPMENT OF PARLIAMENT DURING THE NINETEENTH CENTURY. By G. LOWES DICKINSON, M.A. 8vo, 7s. 6d.

Falkiner.—STUDIES IN IRISH HISTORY AND BIOGRAPHY. Mainly of the Eighteenth Century. By C. LITTON FALKINER. 8vo, 12s. 6d. net.

Fitzmaurice.—CHARLES WILLIAM FERDINAND, Duke of Brunswick; an Historical Study. By Lord EDMOND FITZMAURICE. With Map and 2 Portraits. 8vo, 6s. net.

Froude (JAMES A.).

THE HISTORY OF ENGLAND, from the Fall of Wolsey to the Defeat of the Spanish Armada. 12 vols. Cr. 8vo, 3s. 6d. each.

THE DIVORCE OF CATHERINE OF ARAGON. Crown 8vo, 3s. 6d.

THE SPANISH STORY OF THE ARMADA, and other Essays. Crown 8vo, 3s. 6d.

Froude (JAMES A.)—*continued.*

THE ENGLISH IN IRELAND IN THE EIGHTEENTH CENTURY. 3 vols. Crown 8vo, 10s. 6d.

ENGLISH SEAMEN IN THE SIXTEENTH CENTURY.

Cabinet Edition. Crown 8vo, 6s.

Illustrated Edition. With 5 Photogravure Plates and 16 other Illustrations. Large Crown 8vo, gilt top, 6s. net.

'*Silver Library*' *Edition.* Crown 8vo, 3s. 6d.

THE COUNCIL OF TRENT. Crown 8vo, 3s. 6d.

SHORT STUDIES ON GREAT SUBJECTS.

Cabinet Edition. 4 vols. 21s.

'*Silver Library*' *Edition.* 4 vols. Cr. 8vo, 3s. 6d. each.

CÆSAR: a Sketch. Cr. 8vo, 3s. 6d.

SELECTIONS FROM THE WRITINGS OF JAMES ANTHONY FROUDE. Edited by P. S. ALLEN, M.A. Crown 8vo, 3s. 6d.

Fuller.—EGYPT AND THE HINTERLAND. By FREDERIC W. FULLER. With Frontispiece and Map of Egypt and the Sudan. 8vo, 10s. 6d. net.

Gardiner (SAMUEL RAWSON, D.C.L., LL.D.).

HISTORY OF ENGLAND, from the Accession of James I. to the Outbreak of the Civil War, 1603-1642. 10 vols. Crown 8vo, 5s. net each.

A HISTORY OF THE GREAT CIVIL WAR, 1642-1649. 4 vols. Crown 8vo, 5s. net each.

A HISTORY OF THE COMMONWEALTH AND THE PROTECTORATE. 1649-1660. Vol. I. 1649-1651. With 14 Maps. 8vo, 21s. Vol. II. 1651-1654. With 7 Maps. 8vo, 21s. Vol. III. 1654-1656. With 6 Maps. 8vo, 21s.

THE STUDENT'S HISTORY OF ENGLAND. With 378 Illustrations. Cr. 8vo, gilt top, 12s.

Also in Three Volumes, price 4s. each.

History, Politics, Polity, Political Memoirs, etc.—*continued.*

Gardiner (SAMUEL RAWSON, D.C.L., LL.D.)—*continued.*

WHAT GUNPOWDER PLOT WAS. With 8 Illustrations. Cr. 8vo, 5s.

CROMWELL'S PLACE IN HISTORY. Founded on Six Lectures delivered in the University of Oxford. Crown 8vo, 3s. 6d.

OLIVER CROMWELL. With Frontispiece. Crown 8vo, 5s. net.

German Empire (The) of To-day: Outlines of its Formation and Development. By 'VERITAS'. Crown 8vo, 6s. net.

Graham.—ROMAN AFRICA: an Outline of the History of the Roman Occupation of North Africa, based chiefly upon Inscriptions and Monumental Remains in that country. By ALEXANDER GRAHAM, F.S.A., F.R.I.B.A. With 30 reproductions of Original Drawings by the Author, and 2 Maps. 8vo, 16s. net.

Greville.—A JOURNAL OF THE REIGNS OF KING GEORGE IV., KING WILLIAM IV., AND QUEEN VICTORIA. By CHARLES C. F. GREVILLE, formerly Clerk of the Council. 8 vols. Crown 8vo, 3s. 6d. each.

Gross.—THE SOURCES AND LITERATURE OF ENGLISH HISTORY, from the Earliest Times to about 1485. By CHARLES GROSS, Ph.D. 8vo, 18s. net.

Hamilton.—HISTORICAL RECORD OF THE 14TH (KING'S) HUSSARS, from A.D. 1715 to A.D. 1900. By Colonel HENRY BLACKBURNE HAMILTON, M.A., Christ Church, Oxford; late commanding the Regiment. With 15 Coloured Plates, 35 Portraits, etc., in Photogravure, and 10 Maps and Plans. Crown 4to, gilt edges, 42s. net.

Hill.—LIBERTY DOCUMENTS. With Contemporary Exposition and Critical Comments drawn from various Writers. Selected and Prepared by MABEL HILL. Edited with an Introduction by ALBERT BUSHNELL HART, Ph.D. Large Crown 8vo, 7s. 6d. net.

HARVARD HISTORICAL STUDIES.

THE SUPPRESSION OF THE AFRICAN SLAVE TRADE TO THE UNITED STATES OF AMERICA, 1638-1870. By W. E. B. Du BOIS, Ph.D. 8vo, 7s. 6d.

THE CONTEST OVER THE RATIFICATION OF THE FEDERAL CONSTITUTION IN MASSACHUSETTS. By S. B. HARDING, A.M. 8vo, 6s.

A CRITICAL STUDY OF NULLIFICATION IN SOUTH CAROLINA. By D. F. HOUSTON, A.M. 8vo, 6s.

NOMINATIONS FOR ELECTIVE OFFICE IN THE UNITED STATES. By FREDERICK W. DALLINGER, A.M. 8vo, 7s. 6d.

A BIBLIOGRAPHY OF BRITISH MUNICIPAL HISTORY, including Gilds and Parliamentary Representation. By CHARLES GROSS, Ph.D. 8vo, 12s.

THE LIBERTY AND FREE-SOIL PARTIES IN THE NORTH-WEST. By THEODORE C. SMITH, Ph.D. 8vo, 7s. 6d.

THE PROVINCIAL GOVERNOR IN THE ENGLISH COLONIES OF NORTH AMERICA. By EVARTS BOUTELL GREENE. 8vo, 7s. 6d.

THE COUNTY PALATINE OF DURHAM: a Study in Constitutional History. By GAILLARD THOMAS LAPSLEY, Ph.D. 8vo, 10s. 6d.

THE ANGLICAN EPISCOPATE AND THE AMERICAN COLONIES. By ARTHUR LYON CROSS, Ph.D. 8vo, 10s. 6d.

Historic Towns.—Edited by E. A. FREEMAN, D.C.L., and Rev. WILLIAM HUNT, M.A. With Maps and Plans. Crown 8vo, 3s. 6d. each.

Bristol. By Rev. W. Hunt.

Carlisle. By Mandell Creighton, D.D.

Cinque Ports. By Montagu Burrows.

Colchester. By Rev. E. L. Cutts.

Exeter. By E. A. Freeman.

London. By Rev. W. J. Loftie.

Oxford. By Rev. C. W. Boase.

Winchester. By G. W. Kitchin, D.D.

York. By Rev. James Raine.

New York. By Theodore Roosevelt.

Boston (U.S.). By Henry Cabot Lodge.

History, Politics, Polity, Political Memoirs, etc.—*continued.*

Hunter.—A HISTORY OF BRITISH INDIA. By Sir WILLIAM WILSON HUNTER, K.C.S.I., M.A., LL.D.

Vol. I.—Introductory to the Overthrow of the English in the Spice Archipelago, 1623. With 4 Maps. 8vo, 18s.

Vol. II.—To the Union of the Old and New Companies under the Earl of Godolphin's Award. 1708. 8vo, 16s.

Ingram.—A CRITICAL EXAMINATION OF IRISH HISTORY. From the Elizabethan Conquest to the Legislative Union of 1800. By T. DUNBAR INGRAM, LL.D. 2 vols. 8vo, 24s.

Joyce. — A SHORT HISTORY OF IRELAND, from the Earliest Times to 1603. By P. W. JOYCE, LL.D. Crown 8vo, 10s. 6d.

Kaye and Malleson.—HISTORY OF THE INDIAN MUTINY, 1857-1858. By Sir JOHN W. KAYE and Colonel G. B. MALLESON. With Analytical Index and Maps and Plans. 6 vols. Crown 8vo, 3s. 6d. each.

Kent.—THE ENGLISH RADICALS: an Historical Sketch. By C. B. ROYLANCE KENT. Crown 8vo, 7s. 6d.

Lang (ANDREW).

THE MYSTERY OF MARY STUART. With 6 Photogravure Plates (4 Portraits) and 15 other Illustrations. 8vo, 18s. net.

JAMES THE SIXTH AND THE GOWRIE MYSTERY. With Gowrie's Coat of Arms in colour, 2 Photogravure Portraits and other Illustrations. 8vo, 12s. 6d. net.

Laurie.—HISTORICAL SURVEY OF PRE-CHRISTIAN EDUCATION. By S. S. LAURIE, A.M., LL.D. Crown 8vo, 7s. 6d.

Lecky.—(The Rt. Hon. WILLIAM E. H.). HISTORY OF ENGLAND IN THE EIGHTEENTH CENTURY.

Library Edition. 8 vols. 8vo. Vols. I. and II., 1700-1760, 36s. Vols. III. and IV., 1760-1784, 36s. Vols. V. and VI., 1784-1793, 36s. Vols. VII. and VIII., 1793-1800, 36s.

Cabinet Edition. ENGLAND. 7 vols. Crown 8vo, 5s. net each. IRELAND. 5 vols. Crown 8vo, 5s. net each.

Lecky.—(The Rt. Hon. WILLIAM E. H.) —*continued.*

HISTORY OF EUROPEAN MORALS FROM AUGUSTUS TO CHARLEMAGNE. 2 vols. Crown 8vo, 10s. net.

HISTORY OF THE RISE AND INFLUENCE OF THE SPIRIT OF RATIONALISM IN EUROPE. 2 vols. Crown 8vo, 10s. net.

DEMOCRACY AND LIBERTY.

Library Edition. 2 vols. 8vo, 36s.

Cabinet Edition. 2 vols. Cr. 8vo, 10s. net.

Lowell.—GOVERNMENTS AND PARTIES IN CONTINENTAL EUROPE. By A. LAWRENCE LOWELL. 2 vols. 8vo, 21s.

Lumsden's Horse, Records of.—Edited by H. H. S. PEARSE. With a Map, and numerous Portraits and Illustrations in the Text. 4to.

Lynch.—THE WAR OF THE CIVILISATIONS: BEING A RECORD OF 'A FOREIGN DEVIL'S' EXPERIENCES WITH THE ALLIES IN CHINA. By GEORGE LYNCH, Special Correspondent of the *Sphere*, etc. With Portrait and 21 Illustrations. Crown 8vo, 6s. net.

Macaulay (LORD).

THE LIFE AND WORKS OF LORD MACAULAY.

'*Edinburgh*' *Edition.* 10 vols. 8vo, 6s. each.

Vols. I.-IV. HISTORY OF ENGLAND.

Vols. V.-VII. ESSAYS, BIOGRAPHIES, INDIAN PENAL CODE, CONTRIBUTIONS TO KNIGHT'S 'QUARTERLY MAGAZINE'.

Vol. VIII. SPEECHES, LAYS OF ANCIENT ROME, MISCELLANEOUS POEMS.

Vols. IX. and X. THE LIFE AND LETTERS OF LORD MACAULAY. By Sir G. O. TREVELYAN, Bart.

History, Politics, Polity, Political Memoirs, etc.—*continued.*

Macaulay (LORD)—*continued.*

THE WORKS.

'Albany' Edition. With 12 Portraits. 12 vols. Large Crown 8vo, 3s. 6d. each.

Vols. I.-VI. HISTORY OF ENG-LAND, FROM THE ACCESSION OF JAMES THE SECOND.

Vols. VII.-X. ESSAYS AND BIO-GRAPHIES.

Vols. XI.-XII. SPEECHES, LAYS OF ANCIENT ROME, ETC., AND INDEX.

Cabinet Edition. 16 vols. Post 8vo, £4 16s.

Library Edition. 5 vols. 8vo, £4.

HISTORY OF ENGLAND FROM THE ACCESSION OF JAMES THE SECOND.

Popular Edition. 2 vols. Cr. 8vo, 5s.

Student's Edition. 2 vols. Cr. 8vo, 12s.

People's Edition. 4 vols. Cr. 8vo, 16s.

'Albany' Edition. With 6 Portraits. 6 vols. Large Crown 8vo, 3s. 6d. each.

Cabinet Edition. 8 vols. Post 8vo, 48s.

'Edinburgh' Edition. 4 vols. 8vo, 6s. each.

CRITICAL AND HISTORICAL ESSAYS, WITH LAYS OF ANCIENT ROME, ETC., in 1 volume.

Popular Edition. Crown 8vo, 2s. 6d.

'Silver Library' Edition. With Portrait and 4 Illustrations to the 'Lays'. Crown 8vo, 3s. 6d.

CRITICAL AND HISTORICAL ESSAYS.

Student's Edition. 1 vol. Cr. 8vo, 6s.

People's Edition. 2 vols. Crown 8vo, 8s.

'Trevelyan' Edition. 2 vols. Crown 8vo, 9s.

Cabinet Edition. 4 vols. Post 8vo. 24s.

'Edinburgh' Edition. 3 vols. 8vo, 6s. each.

Library Edition. 3 vols. 8vo, 36s.

Macaulay (LORD)—*continued.*

ESSAYS, which may be had separately, sewed, 6d. each ; cloth, 1s. each.

Addison and Walpole.
Croker's Boswell's Johnson.
Hallam's Constitutional History.
Warren Hastings.
The Earl of Chatham (Two Essays).

Frederic the Great.
Ranke and Gladstone.
Lord Bacon.
Lord Clive.
Lord Byron, and The Comic Dramatists of the Restoration.

MISCELLANEOUS WRITINGS, SPEECHES, AND POEMS.

Popular Edition. Crown 8vo, 2s. 6d.

Cabinet Edition. 4 vols. Post 8vo, 24s.

SELECTIONS FROM THE WRITINGS OF LORD MACAULAY. Edited, with Occasional Notes, by the Right Hon. Sir G. O. TREVELYAN, Bart. Crown 8vo, 6s.

Mackinnon (JAMES, Ph.D.).

THE HISTORY OF EDWARD THE THIRD. 8vo, 18s.

THE GROWTH AND DECLINE OF THE FRENCH MONARCHY. 8vo, 21s. net.

Mallet.—MALLET DU PAN AND THE FRENCH REVOLUTION. By BERNARD MALLET. With Photogravure Portrait. 8vo, 12s. 6d.

May.—THE CONSTITUTIONAL HIS-TORY OF ENGLAND since the Accession of George III. 1760-1870. By Sir THOMAS ERSKINE MAY, K.C.B. (Lord Farnborough). 3 vols. Cr. 8vo, 18s.

Merivale (CHARLES, D.D.).

HISTORY OF THE ROMANS UNDER THE EMPIRE. 8 vols. Crown 8vo, 3s. 6d. each.

THE FALL OF THE ROMAN RE-PUBLIC: a Short History of the Last Century of the Commonwealth. 12mo, 7s. 6d.

GENERAL HISTORY OF ROME, from the Foundation of the City to the Fall of Augustulus, B.C. 753-A.D. 476. With 5 Maps. Cr. 8vo, 7s. 6d.

Montague.—THE ELEMENTS OF ENGLISH CONSTITUTIONAL HIS-TORY. By F. C. MONTAGUE, M.A. Crown 8vo, 3s. 6d.

Nash.—THE GREAT FAMINE AND ITS CAUSES. By VAUGHAN NASH. With 8 Illustrations from Photographs by the Author, and a Map of India showing the Famine Area. Cr. 8vo, 6s.

History, Politics, Polity, Political Memoirs, etc.—*continued.*

Owens College Essays.—Edited by T. F. TOUT, M.A., Professor of History in the Owens College, Victoria University, and JAMES TAIT, M.A., Assistant Lecturer in History. With 4 Maps. 8vo, 12s. 6d. net.

Powell and Trevelyan.—THE PEASANTS' RISING AND THE LOLLARDS: a Collection of Unpublished Documents. Edited by EDGAR POWELL and G. M. TREVELYAN. 8vo, 6s. net.

Randolph.—THE LAW AND POLICY OF ANNEXATION, with Special Reference to the Philippines; together with Observations on the Status of Cuba. By CARMAN F. RANDOLPH. 8vo, 9s. net.

Rankin (REGINALD).
THE MARQUIS D'ARGENSON AND RICHARD THE SECOND. 8vo, 10s. 6d. net.
A SUBALTERN'S LETTERS TO HIS WIFE. (The Boer War.) Cr. 8vo, 3s. 6d.

Ransome.—THE RISE OF CONSTITUTIONAL GOVERNMENT IN ENGLAND. By CYRIL RANSOME, M.A. Crown 8vo, 6s.

Seebohm (FREDERIC, LL.D., F.S.A.).
THE ENGLISH VILLAGE COMMUNITY. With 13 Maps and Plates. 8vo, 16s.
TRIBAL CUSTOM IN ANGLO-SAXON LAW: being an Essay supplemental to (1) 'The English Village Community,' (2) 'The Tribal System in Wales'. 8vo, 16s.

Seton-Karr.—THE CALL TO ARMS, 1900-1901; or a Review of the Imperial Yeomaury Movement, and some subjects connected therewith. By Sir HENRY SETON-KARR, M.P. With a Frontispiece by R. CATON WOODVILLE. Cr. 8vo, 5s. net.

Shaw.—A HISTORY OF THE ENGLISH CHURCH DURING THE CIVIL WARS AND UNDER THE COMMONWEALTH, 1640-1660. By WM. A. SHAW, Litt.D. 2 vols. 8vo, 36s.

Sheppard.—THE OLD ROYAL PALACE OF WHITEHALL. By EDGAR SHEPPARD, D.D., Sub-Dean of H.M. Chapels Royal, Sub-Almoner to the King. With 6 Photogravure Plates and 33 other Illustrations. Medium 8vo, 21s. net.

Smith.—CARTHAGE AND THE CARTHAGINIANS. By R. BOSWORTH SMITH, M.A. With Maps, Plans, etc. Crown 8vo, 3s. 6d.

Stephens.—A HISTORY OF THE FRENCH REVOLUTION. By H. MORSE STEPHENS. 8vo. Vols. I. and II. 18s. each.

Sternberg.—MY EXPERIENCES OF THE BOER WAR. By ADALBERT COUNT STERNBERG. With Preface by Lieut.-Col. G. F. R. HENDERSON. Cr. 8vo, 5s. net.

Stubbs.—HISTORY OF THE UNIVERSITY OF DUBLIN. By J. W. STUBBS. 8vo, 12s. 6d.

Stubbs.—HISTORICAL INTRODUCTION TO THE 'ROLLS SERIES'. By WILLIAM STUBBS, D.D., formerly Bishop of Oxford and Regius Professor of Modern History in the University. Collected and Edited by ARTHUR HASSALL, M.A. 8vo, 12s. 6d. net.

Sutherland. THE HISTORY OF AUSTRALIA AND NEW ZEALAND, from 1606-1900. By ALEXANDER SUTHERLAND, M.A., and GEORGE SUTHERLAND, M.A. Crown 8vo, 2s. 6d.

Taylor.—A STUDENT'S MANUAL OF THE HISTORY OF INDIA. By Colonel MEADOWS TAYLOR, C.S.I., etc. Crown 8vo, 7s. 6d.

Thomson. CHINA AND THE POWERS: a Narrative of the Outbreak of 1900. By H. C. THOMSON. With 2 Maps and 29 Illustrations. 8vo, 10s. 6d. net.

Todd. PARLIAMENTARY GOVERNMENT IN THE BRITISH COLONIES. By ALPHEUS TODD, LL.D. 8vo, 30s. net.

Trevelyan. THE AMERICAN REVOLUTION. Part I. 1766-1776. By Sir G. O. TREVELYAN, Bart. 8vo, 16s.

Trevelyan. ENGLAND IN THE AGE OF WYCLIFFE. By GEORGE MACAULAY TREVELYAN. 8vo, 15s.

History, Politics, Polity, Political Memoirs, etc.—*continued.*

Wakeman and Hassall.—ESSAYS INTRODUCTORY TO THE STUDY OF ENGLISH CONSTITUTIONAL HISTORY. Edited by HENRY OFFLEY WAKEMAN, M.A., and ARTHUR HASSALL, M.A. Crown 8vo, 6s.

Walpole.—HISTORY OF ENGLAND FROM THE CONCLUSION OF THE GREAT WAR IN 1815 TO 1858. By Sir SPENCER WALPOLE, K.C.B. 6 vols. Crown 8vo, 6s. each.

Wylie (JAMES HAMILTON, M.A.).

HISTORY OF ENGLAND UNDER HENRY IV. 4 vols. Crown 8vo. Vol. I., 1399-1404, 10s. 6d. Vol. II., 1405-1406, 15s. (*out of print*). Vol. III., 1407-1411, 15s. Vol. IV., 1411-1413, 21s.

THE COUNCIL OF CONSTANCE TO THE DEATH OF JOHN HUS. Cr. 8vo, 6s. net.

Biography, Personal Memoirs, etc.

Bacon.—THE LETTERS AND LIFE OF FRANCIS BACON, INCLUDING ALL HIS OCCASIONAL WORKS. Edited by JAMES SPEDDING. 7 vols. 8vo, £4 4s.

Bagehot. — BIOGRAPHICAL STUDIES. By WALTER BAGEHOT. Crown 8vo, 3s. 6d.

BERNARDS (THE) OF ABINGTON AND NETHER WINCHENDON: a Family History. By Mrs. NAPIER HIGGINS. 2 vols. 8vo.

Blount.—THE MEMOIRS OF SIR EDWARD BLOUNT, K.C.B., etc. Edited by STUART J. REID, Author of 'The Life and Times of Sydney Smith,' etc. With 3 Photogravure Plates. 8vo, 10s. 6d. net.

Bowen. — EDWARD BOWEN: A MEMOIR. By the Rev. the Hon. W. E. BOWEN. With Appendices, 3 Photogravure Portraits and 2 other Illustrations. 8vo, 12s. 6d. net.

Carlyle.—THOMAS CARLYLE: A History of his Life. By JAMES ANTHONY FROUDE. Crown 8vo. 1795-1835. 2 vols. 7s. 1834-1881. 2 vols. 7s.

Crozier.—MY INNER LIFE: being a Chapter in Personal Evolution and Autobiography. By JOHN BEATTIE CROZIER, LL.D. 8vo, 14s.

Dante.—THE LIFE AND WORKS OF DANTE ALLIGHIERI: being an Introduction to the Study of the 'Divina Commedia'. By the Rev. J. F. HOGAN, D.D. With Portrait. 8vo, 12s. 6d.

Danton.—LIFE OF DANTON. By A. H. BEESLY. With Portraits. Cr. 8vo, 6s.

De Bode. — THE BARONESS DE BODE, 1775-1803. By WILLIAM S. CHILDE-PEMBERTON. With 4 Photogravure Portraits and other Illustrations. 8vo, gilt top, 12s. 6d. net.

Erasmus.

LIFE AND LETTERS OF ERASMUS. By JAMES ANTHONY FROUDE. Crown 8vo, 3s. 6d.

THE EPISTLES OF ERASMUS, from his earliest Letters to his Fifty-first Year, arranged in Order of Time. English Translations, with a Commentary. By FRANCIS MORGAN NICHOLS. 8vo, 18s. net.

Faraday.—FARADAY AS A DISCOVERER. By JOHN TYNDALL. Crown 8vo, 3s. 6d.

Fénelon: his Friends and his Enemies, 1651-1715. By E. K. SANDERS. With Portrait. 8vo, 10s. 6d.

Fox.—THE EARLY HISTORY OF CHARLES JAMES FOX. By the Right Hon. Sir G. O. TREVELYAN, Bart. Crown 8vo, 3s. 6d.

Froude.—HURRELL FROUDE: Some Reprints and Reprinted Comments. With a Biographical Preface by the Editor, L. I. GUINEY. With Illustrations. 8vo.

Granville. — SOME RECORDS OF THE LATER LIFE OF HARRIET, COUNTESS GRANVILLE. By her Grand-daughter, the Hon. MRS. OLDFIELD. With 17 Portraits. 8vo, gilt top, 16s. net.

Grey.—MEMOIR OF SIR GEORGE GREY, BART., G.C.B., 1799-1882. By MANDELL CREIGHTON, D.D., late Lord Bishop of London. With 3 Portraits. Crown 8vo, 6s. net.

Hamilton.—LIFE OF SIR WILLIAM HAMILTON. By R. P. GRAVES. 8vo, 3 vols. 15s. each. ADDENDUM. 8vo, 6d. sewed.

Harrow School Register (The), 1801 - 1900. Second Edition, 1901. Edited by M. G. DAUGLISH, Barrister-at-Law. 8vo, 15s. net.

Biography, Personal Memoirs, etc.—*continued.*

Havelock. — MEMOIRS OF SIR HENRY HAVELOCK, K.C.B. By JOHN CLARK MARSHMAN. Cr. 8vo, 3s. 6d.

Haweis.—MY MUSICAL LIFE. By the Rev. H. R. HAWEIS. With Portrait of Richard Wagner and 3 Illustrations. Crown 8vo, 6s. net.

Hunter. — THE LIFE OF SIR WILLIAM WILSON HUNTER, K.C.S.I., M.A., LL.D. Author of 'A History of British India,' etc. By FRANCIS HENRY SKRINE, F.S.S. With 6 Portraits (2 Photogravures) and 4 other Illustrations. 8vo, 16s. net.

Jackson.—STONEWALL JACKSON AND THE AMERICAN CIVIL WAR. By Lieut.-Col. G. F. R. HENDERSON. With 2 Portraits and 33 Maps and Plans. 2 vols. Crown 8vo, 16s. net.

Kielmansegge. — DIARY OF A JOURNEY TO ENGLAND IN THE YEARS 1761-1762. By Count FREDERICK KIELMANSEGGE. With 4 Illustrations. Crown 8vo, 5s. net.

Lieven.—LETTERS OF DOROTHEA, PRINCESS LIEVEN, DURING HER RESIDENCE IN LONDON, 1812-1834. Edited by LIONEL G. ROBINSON. With 2 Photogravure Portraits. 8vo, 14s. net.

Luther.—LIFE OF LUTHER. By JULIUS KÖSTLIN. With 62 Illustrations and 4 Facsimiles of MSS. Crown 8vo, 3s. 6d.

Macaulay.—THE LIFE AND LETTERS OF LORD MACAULAY. By the Right Hon. Sir G. O. TREVELYAN, Bart.
Popular Edition. 1 vol. Cr. 8vo, 2s. 6d.
Student's Edition. 1 vol. Cr. 8vo. 6s.
Cabinet Edition. 2 vols. Post 8vo, 12s.
'Edinburgh' Edition. 2 vols. 8vo, 6s. each.
Library Edition. 2 vols. 8vo, 36s.

Max Muller (F.)
THE LIFE AND LETTERS OF THE RIGHT HON. FRIEDRICH MAX MÜLLER. Edited by his Wife. With Photogravure Portraits and other Illustrations. 2 vols. 8vo, 32s. net.
MY AUTOBIOGRAPHY: a Fragment. With 6 Portraits. 8vo, 12s. 6d.
AULD LANG SYNE. Second Series. 8vo, 10s. 6d.
CHIPS FROM A GERMAN WORKSHOP. Vol. II. Biographical Essays. Crown 8vo, 5s.

Meade.—GENERAL SIR RICHARD MEADE AND THE FEUDATORY STATES OF CENTRAL AND SOUTHERN INDIA. By THOMAS HENRY THORNTON. With Portrait, Map and Illustrations. 8vo, 10s. 6d. net.

Morris.—THE LIFE OF WILLIAM MORRIS. By J. W. MACKAIL. With 2 Portraits and 8 other Illustrations by E. H. NEW, etc. 2 vols. Large Crown 8vo, 10s. net.

On the Banks of the Seine.—By A. M. F., Authoress of 'Foreign Courts and Foreign Homes'. Crown 8vo, 6s.

Paget.—MEMOIRS AND LETTERS OF SIR JAMES PAGET. Edited by STEPHEN PAGET, one of his sons. With 6 Portraits (3 Photogravures) and 4 other Illustrations. 8vo, 12s. 6d. net.

Place.—THE LIFE OF FRANCIS PLACE, 1771-1854. By GRAHAM WALLAS, M.A. With 2 Portraits. 8vo, 12s.

Powys.—PASSAGES FROM THE DIARIES OF MRS. PHILIP LYBBE POWYS, OF HARDWICK HOUSE, OXON. 1756-1808. Edited by EMILY J. CLIMENSON. 8vo, gilt top, 16s.

Râmakrishna : his Life and Sayings. By the Right Hon. F. MAX MÜLLER. Crown 8vo, 5s.

Rich. — MARY RICH, COUNTESS OF WARWICK (1625-1678) : Her Family and Friends. By C. FELL SMITH. With 7 Photogravure Portraits and 9 other Illustrations. 8vo, gilt top, 18s. net.

Rochester and other Literary Rakes of the Court of Charles II., with some Account of their Surroundings. By the Author of 'The Life of Sir Kenelm Digby,' 'The Life of a Prig,' etc. With 15 Portraits. 8vo, 16s.

Romanes.—THE LIFE AND LETTERS OF GEORGE JOHN ROMANES, M.A., LL.D., F.R.S. Written and Edited by his WIFE. With Portrait and 2 Illustrations. Crown 8vo, 5s. net.

Russell. — SWALLOWFIELD AND ITS OWNERS. By CONSTANCE, Lady RUSSELL of Swallowfield Park. With 15 Photogravure Portraits and 36 other Illustrations. 4to, gilt edges, 42s. net.

Biography, Personal Memoirs, etc.—*continued.*

Seebohm.—THE OXFORD REFORMERS—JOHN COLET, ERASMUS, AND THOMAS MORE : a History of their Fellow-Work. By FREDERIC SEEBOHM. 8vo, 14s.

Shakespeare.—OUTLINES OF THE LIFE OF SHAKESPEARE. By J. O. HALLIWELL-PHILLIPPS. With Illustrations and Facsimiles. 2 vols. Royal 8vo, 21s.

Tales of my Father.—By A. M. F. Author of 'Foreign Courts and Foreign Homes,' and 'On the Banks of the Seine'. Crown 8vo, 6s.

Tallentyre.—THE WOMEN OF THE SALONS, and other French Portraits. By S. G. TALLENTYRE. With 11 Photogravure Portraits. 8vo, 10s. 6d. net.

Victoria, Queen, 1819-1901. By RICHARD R. HOLMES, M.V.O., F.S.A. Librarian to the Queen. With Photogravure Portrait. Cr. 8vo, gilt top, 5s. net.

Walpole. — SOME UNPUBLISHED LETTERS OF HORACE WALPOLE. Edited by Sir SPENCER WALPOLE, K.C.B. With 2 Portraits. Cr. 8vo, 4s. 6d. net.

Wellington.—LIFE OF THE DUKE OF WELLINGTON. By the Rev. G. R. GLEIG, M.A. Crown 8vo, 3s. 6d.

Travel and Adventure, the Colonies, etc.

Arnold.—SEAS AND LANDS. By Sir EDWIN ARNOLD. With 71 Illustrations. Crown 8vo, 3s. 6d.

Baker (Sir S. W.).

EIGHT YEARS IN CEYLON. With 6 Illustrations. Crown 8vo, 3s. 6d.

THE RIFLE AND THE HOUND IN CEYLON. With 6 Illustrations. Crown 8vo, 3s. 6d.

Ball (JOHN).

THE ALPINE GUIDE. Reconstructed and Revised on behalf of the Alpine Club by W. A. B. COOLIDGE.

Vol. I., THE WESTERN ALPS : the Alpine Region, South of the Rhone Valley, from the Col de Tenda to the Simplon Pass. With 9 New and Revised Maps. Crown 8vo, 12s. net.

HINTS AND NOTES, PRACTICAL AND SCIENTIFIC, FOR TRAVELLERS IN THE ALPS: being a revision of the General Introduction to the 'Alpine Guide'. Crown 8vo, 3s. net.

Bent.—THE RUINED CITIES OF MASHONALAND : being a Record of Excavation and Exploration in 1891. By J. THEODORE BENT. With 117 Illustrations. Crown 8vo, 3s. 6d.

Brassey (THE LATE LADY).

A VOYAGE IN THE 'SUNBEAM'; OUR HOME ON THE OCEAN FOR ELEVEN MONTHS.

Cabinet Edition. With Map and 66 Illustrations. Crown 8vo, gilt edges, 7s. 6d.

'*Silver Library*' *Edition.* With 66 Illustrations. Crown 8vo, 3s. 6d.

Popular Edition. With 60 Illustrations. 4to, 6d. sewed, 1s. cloth.

School Edition. With 37 Illustrations. Fcp., 2s. cloth, or 3s. white parchment.

SUNSHINE AND STORM IN THE EAST.

Popular Edition. With 103 Illustrations. 4to, 6d. sewed, 1s. cloth.

IN THE TRADES, THE TROPICS, AND THE 'ROARING FORTIES'.

Cabinet Edition. With Map and 220 Illustrations. Crown 8vo, gilt edges, 7s. 6d.

Fountain (PAUL).

THE GREAT DESERTS AND FORESTS OF NORTH AMERICA. With a Preface by W. H. HUDSON, Author of 'The Naturalist in La Plata,' etc. 8vo, 9s. 6d. net.

THE GREAT MOUNTAINS AND FORESTS OF SOUTH AMERICA. With Portrait and 7 Illustrations. 8vo, 10s. 6d.

Travel and Adventure, the Colonies, etc.—*continued.*

Froude (JAMES A.).

OCEANA : or England and her Colonies. With 9 Illustrations. Crown 8vo, 3s. 6d.

THE ENGLISH IN THE WEST INDIES : or, the Bow of Ulysses. With 9 Illustrations. Crown 8vo, 2s. boards, 2s. 6d. cloth.

Grove. — SEVENTY - ONE DAYS' CAMPING IN MOROCCO. By Lady GROVE. With Photogravure Portrait and 32 Illustrations from Photographs. 8vo, 7s. 6d. net.

Haggard.—A WINTER PILGRIMAGE : Being an Account of Travels through Palestine, Italy and the Island of Cyprus, undertaken in the year 1900. By H. RIDER HAGGARD. With 31 Illustrations from Photographs. Crown 8vo, gilt top, 12s. 6d. net.

Heathcote.—ST. KILDA. By NORMAN HEATHCOTE. With 80 Illustrations from Sketches and Photographs of the People, Scenery and Birds, by the Author. 8vo, 10s. 6d. net.

Howitt.—VISITS TO REMARKABLE PLACES. Old Halls, Battlefields, Scenes, illustrative of Striking Passages in English History and Poetry. By WILLIAM HOWITT. With 80 Illustrations. Crown 8vo, 3s. 6d.

Knight (E. F.).

WITH THE ROYAL TOUR : a Narrative of the Recent Tour of the Duke and Duchess of Cornwall and York through Greater Britain. With 16 Illustrations and a Map. Crown 8vo, 5s. net.

THE CRUISE OF THE 'ALERTE': the Narrative of a search for Treasure on the Desert Island of Trinidad. With 2 Maps and 23 Illustrations. Crown 8vo, 3s. 6d.

WHERE THREE EMPIRES MEET : a Narrative of Recent Travel in Kashmir, Western Tibet, Baltistan, Ladak, Gilgit, and the adjoining Countries. With a Map and 54 Illustrations. Crown 8vo, 3s. 6d.

THE 'FALCON' ON THE BALTIC : a Voyage from London to Copenhagen in a Three-Tonner. With 10 Full-page Illustrations. Cr. 8vo, 3s. 6d.

Lees.—PEAKS AND PINES : another Norway Book. By J. A. LEES. With 63 Illustrations and Photographs. Cr. 8vo, 6s.

Lees and Clutterbuck.—B.C. 1887 : A RAMBLE IN BRITISH COLUMBIA. By J. A. LEES and W. J. CLUTTERBUCK. With Map and 75 Illustrations. Crown 8vo, 3s. 6d.

Lynch. — ARMENIA : Travels and Studies. By H. F. B. LYNCH. With 197 Illustrations (some in tints) reproduced from Photographs and Sketches by the Author, 16 Maps and Plans, a Bibliography, and a Map of Armenia and adjacent countries. 2 vols. Medium 8vo, gilt top, 42s. net.

Nansen.—THE FIRST CROSSING OF GREENLAND. By FRIDTJOF NANSEN. With 143 Illustrations and a Map. Cr. 8vo, 3s. 6d.

Rice.—OCCASIONAL ESSAYS ON NATIVE SOUTH INDIAN LIFE. By STANLEY P. RICE, Indian Civil Service. 8vo, 10s. 6d.

Smith.—CLIMBING IN THE BRITISH ISLES. By W. P. HASKETT SMITH. With Illustrations and numerous Plans. Part I. ENGLAND. 16mo, 3s. net. Part II. WALES AND IRELAND. 16mo, 3s. net.

Spender.—TWO WINTERS IN NORWAY : being an Account of Two Holidays spent on Snow-shoes and in Sleigh Driving, and including an Expedition to the Lapps. By A. EDMUND SPENDER. With 40 Illustrations from Photographs. 8vo, 10s. 6d. net.

Stephen. THE PLAYGROUND OF EUROPE (The Alps). By Sir LESLIE STEPHEN, K.C.B. With 4 Illustrations. Cr. 8vo, 3s. 6d.

Three in Norway. By Two of them. With a Map and 59 Illustrations. Cr. 8vo, 2s. boards, 2s. 6d. cloth.

Tyndall (JOHN).

THE GLACIERS OF THE ALPS : With 61 Illustrations. Crown 8vo, 6s. 6d. net.

HOURS OF EXERCISE IN THE ALPS. With 7 Illustrations. Cr. 8vo, 6s. 6d. net.

Willard. - THE LAND OF THE LATINS. By ASHTON R. WILLARD. With 11 Illustrations from Photographs. Crown 8vo, 5s. net.

Sport and Pastime.
THE BADMINTON LIBRARY.

Edited by HIS GRACE THE (EIGHTH) DUKE OF BEAUFORT, K.G., and
A. E. T. WATSON.

ARCHERY. By C. J. LONGMAN, Col. H. WALROND, Miss LEGH, etc. With 2 Maps, 23 Plates, and 172 Illustrations in the Text. Crown 8vo, cloth, 6s. net ; half-bound, with gilt top, 9s. net.

ATHLETICS. By MONTAGUE SHEARMAN, W. BEACHER THOMAS, W. RYE, etc. With 12 Plates and 37 Illustrations in the Text. Crown 8vo, cloth, 6s. net ; half-bound, with gilt top, 9s. net.

BIG GAME SHOOTING. By CLIVE PHILLIPPS-WOLLEY.

Vol. I. AFRICA AND AMERICA. With Contributions by Sir SAMUEL W. BAKER, W. C. OSWELL, F. C. SELOUS, etc. With 20 Plates and 57 Illustrations in the Text. Crown 8vo, cloth, 6s. net ; half-bound, with gilt top, 9s. net.

Vol. II. EUROPE, ASIA, AND THE ARCTIC REGIONS. With Contributions by Lieut.-Colonel R. HEBER PERCY, Major ALGERNON C. HEBER PERCY, etc. With 17 Plates and 56 Illustrations in the Text. Crown 8vo, cloth, 6s. net ; half-bound, with gilt top, 9s. net.

BILLIARDS. By Major W. BROADFOOT, R.E., A. H. BOYD, W. J. FORD, etc. With 11 Plates, 19 Illustrations in the Text, and numerous Diagrams. Crown 8vo, cloth, 6s. net ; half-bound, with gilt top, 9s. net.

COURSING AND FALCONRY. By HARDING COX, CHARLES RICHARDSON, etc. With 20 Plates and 55 Illustrations in the Text. Crown 8vo, cloth, 6s. net ; half-bound, with gilt top, 9s. net.

CRICKET. By A. G. STEEL, the Hon. R. H. LYTTELTON, A. LANG, W. G. GRACE, etc. With 13 Plates and 52 Illustrations in the Text. Crown 8vo, cloth, 6s. net ; half-bound, with gilt top, 9s. net.

CYCLING. By the EARL OF ALBEMARLE and G. LACY HILLIER. With 19 Plates and 44 Illustrations in the Text. Crown 8vo, cloth, 6s. net ; half-bound, with gilt top, 9s. net.

DANCING. By Mrs. LILLY GROVE, etc. With Musical Examples, and 38 Full-page Plates and 93 Illustrations in the Text. Cr. 8vo, cloth, 6s. net ; half-bound, with gilt top, 9s. net.

DRIVING. By His Grace the (Eighth) DUKE OF BEAUFORT, K.G., A. E. T. WATSON, etc. With 12 Plates and 54 Illustrations in the Text. Crown 8vo, cloth, 6s. net ; half-bound, with gilt top, 9s. net.

FENCING, BOXING AND WRESTLING. By WALTER H. POLLOCK, F. C. GROVE, etc. With 18 Plates and 24 Illustrations in the Text. Crown 8vo, cloth, 6s. net ; half-bound, with gilt top, 9s. net.

FISHING. By H. CHOLMONDELEY-PENNELL.

Vol. I.—SALMON AND TROUT. With Contributions by H. R. FRANCIS, Major JOHN P. TRAHERNE, etc. With 9 Plates and numerous Illustrations of Tackle, etc. Crown 8vo, 6s. net ; half-bound, with gilt top, 9s. net.

Vol. II.—PIKE AND OTHER COARSE FISH. With Contributions by WILLIAM SENIOR, G. CHRISTOPHER DAVIS, etc. With 7 Plates and numerous Illustrations of Tackle, etc. Cr. 8vo, cloth, 6s. net ; half-bound, with gilt top, 9s. net.

FOOTBALL.—By MONTAGUE SHEARMAN, W. J. OAKLEY, FRANK MITCHELL, etc. With 19 Plates and 35 Illustrations in the Text. Crown 8vo, cloth, 6s. net ; half-bound, with gilt top, 9s. net.

GOLF. By HORACE G. HUTCHINSON, the Rt. Hon. A. J. BALFOUR, M.P., ANDREW LANG, etc. With 34 Plates and 56 Illustrations in the Text. Crown 8vo, cloth, 6s. net ; half-bound, with gilt top, 9s. net.

Sport and Pastime—*continued.*
THE BADMINTON LIBRARY—*continued.*
Edited by HIS GRACE THE (EIGHTH) DUKE OF BEAUFORT, K.G., and
A. E. T. WATSON.

HUNTING. By His Grace the (Eighth) DUKE OF BEAUFORT, K.G., MOWBRAY MORRIS, G. H. LONGMAN, etc. With 5 Plates and 54 Illustrations in the Text. Crown 8vo, cloth, 6s. net; half-bound, with gilt top, 9s. net.

MOTORS AND MOTOR-DRIVING. By ALFRED C. HARMSWORTH, the Hon. JOHN SCOTT-MONTAGU, etc. With 13 Plates and 136 Illustrations in the Text. Crown 8vo, cloth, 9s. net; half-bound, 12s. net. A Cloth Box for use when Motoring, price 2s. net.

MOUNTAINEERING. By C. T. DENT, the Right Hon. J. BRYCE, M.P., Sir MARTIN CONWAY, etc. With 13 Plates and 91 Illustrations in the Text. Crown 8vo, cloth, 6s. net; half-bound, with gilt top, 9s. net.

POETRY OF SPORT (THE). Selected by HEDLEY PEEK. With 32 Plates and 74 Illustrations in the Text. Cr. 8vo, cloth, 6s. net; half-bound, with gilt top, 9s. net.

RACING AND STEEPLE-CHASING. By the EARL OF SUFFOLK AND BERKSHIRE, W. G. CRAVEN, the Hon. F. LAWLEY, etc. With Frontispiece and 56 Illustrations in the Text. Cr. 8vo, cloth, 6s. net; half-bound, with gilt top, 9s. net.

RIDING AND POLO. By Captain ROBERT WEIR, J. MORAY BROWN, T. F. DALE, the late DUKE OF BEAUFORT, etc. With 18 Plates and 41 Illustrations in the Text. Crown 8vo, cloth, 6s. net; half-bound, with gilt top, 9s. net.

ROWING. By R. P. P. ROWE and C. M. PITMAN, etc. With 75 Illustrations. Crown 8vo, cloth, 6s. net; half-bound, with gilt top, 9s. net.

SEA FISHING. By JOHN BICKERDYKE, Sir H. W. GORE-BOOTH, ALFRED C. HARMSWORTH, and W. SENIOR. With 22 Full-page Plates and 175 Illustrations in the Text. Crown 8vo, cloth, 6s. net; half-bound, with gilt top, 9s. net.

SHOOTING.
Vol. I.—FIELD AND COVERT. By LORD WALSINGHAM, Sir RALPH PAYNE-GALLWEY, Bart., etc. With 11 Plates and 95 Illustrations in the Text. Crown 8vo, cloth, 6s. net; half-bound, with gilt top, 9s. net.
Vol. II.—MOOR AND MARSH. By LORD WALSINGHAM, Sir RALPH PAYNE-GALLWEY, Bart., etc. With 8 Plates and 57 Illustrations in the Text. Crown 8vo, cloth, 6s. net; half-bound, with gilt top, 9s. net.

SKATING, CURLING, TOBOGGANING. By J. M. HEATHCOTE, C. G. TEBBUTT, T. MAXWELL WITHAM, etc. With 12 Plates and 272 Illustrations in the Text. Crown 8vo, cloth, 6s. net; half-bound, with gilt top, 9s. net.

SWIMMING. By ARCHIBALD SINCLAIR and WILLIAM HENRY. With 13 Plates and 112 Illustrations in the Text. Cr. 8vo, cloth, 6s. net; half-bound, with gilt top, 9s. net.

TENNIS, LAWN TENNIS, RACKETS AND FIVES. By J. M. and C. G. HEATHCOTE, E. O. PLEYDELL-BOUVERIE, the Hon. A. LYTTELTON, etc. With 12 Plates and 67 Illustrations in the Text. Crown 8vo, cloth, 6s. net; half-bound, with gilt top, 9s. net.

YACHTING.
Vol. I.—CRUISING, CONSTRUCTION OF YACHTS, YACHT RACING RULES, FITTING-OUT, etc. By Sir EDWARD SULLIVAN, Bart., the EARL OF PEMBROKE, LORD BRASSEY, K.C.B., etc. With 21 Plates and 93 Illustrations in the Text. Crown 8vo, cloth, 6s. net; half-bound, with gilt top, 9s. net.
Vol. II.—YACHT CLUBS, YACHTING IN AMERICA AND THE COLONIES, YACHT RACING, etc. By R. T. PRITCHETT, the MARQUIS OF DUFFERIN AND AVA, K.P., etc. With 35 Plates and 160 Illustrations in the Text. Crown 8vo, cloth, 6s. net; half-bound, with gilt top, 9s. net.

Sport and Pastime—*continued.*

FUR, FEATHER, AND FIN SERIES.

Edited by A. E. T. WATSON.

Crown 8vo, price 5s. each Volume, cloth.

_**_* *The Volumes are also issued half-bound in Leather, with gilt top. Price 7s. 6d. net each.*

THE PARTRIDGE. NATURAL HISTORY, by the Rev. H. A. MACPHERSON; SHOOTING, by A. J. STUART-WORTLEY; COOKERY, by GEORGE SAINTSBURY. With 11 Illustrations and various Diagrams. Crown 8vo, 5s.

THE GROUSE. NATURAL HISTORY, by the Rev. H. A. MACPHERSON; SHOOTING, by A. J. STUART-WORTLEY; COOKERY, by GEORGE SAINTSBURY. With 13 Illustrations and various Diagrams. Crown 8vo, 5s.

THE PHEASANT. NATURAL HISTORY, by the Rev. H. A. MACPHERSON; SHOOTING, by A. J. STUART-WORTLEY; COOKERY, by ALEXANDER INNES SHAND. With 10 Illustrations and various Diagrams. Crown 8vo, 5s.

THE HARE. NATURAL HISTORY, by the Rev. H. A. MACPHERSON; SHOOTING, by the Hon. GERALD LASCELLES; COURSING, by CHARLES RICHARDSON; HUNTING, by J. S. GIBBONS and G. H. LONGMAN; COOKERY, by Col. KENNEY HERBERT. With 9 Illustrations. Crown 8vo, 5s.

RED DEER. NATURAL HISTORY, by the Rev. H. A. MACPHERSON; DEER STALKING, by CAMERON OF LOCHIEL; STAG HUNTING, by Viscount EBRINGTON; COOKERY, by ALEXANDER INNES SHAND. With 10 Illustrations. Crown 8vo, 5s.

THE SALMON. By the Hon. A. E. GATHORNE-HARDY. With Chapters on the Law of Salmon Fishing by CLAUD DOUGLAS PENNANT; COOKERY, by ALEXANDER INNES SHAND. With 8 Illustrations. Crown 8vo, 5s.

THE TROUT. By the MARQUESS OF GRANBY. With Chapters on the Breeding of Trout by Col. H. CUSTANCE; and COOKERY, by ALEXANDER INNES SHAND. With 12 Illustrations. Crown 8vo, 5s.

THE RABBIT. By JAMES EDMUND HARTING. COOKERY, by ALEXANDER INNES SHAND. With 10 Illustrations. Crown 8vo, 5s.

PIKE AND PERCH. By WILLIAM SENIOR ('Redspinner,' Editor of the *Field*). With Chapters by JOHN BICKERDYKE and W. H. POPE. COOKERY, by ALEXANDER INNES SHAND. With 12 Illustrations. Crown 8vo, 5s.

Alverstone and Alcock.—SURREY CRICKET: Its History and Associations. Edited by the Right Hon. LORD ALVERSTONE, L.C.J., President, and C. W. ALCOCK, Secretary, of the Surrey County Cricket Club. With 48 Illustrations. 8vo, 16s. net.

Bickerdyke.—DAYS OF MY LIFE ON WATER, FRESH AND SALT: and other papers. By JOHN BICKERDYKE. With Photo-Etching Frontispiece and 8 Full-page Illustrations. Crown 8vo, 3s. 6d.

Blackburne.—MR. BLACKBURNE'S GAMES AT CHESS. Selected, Annotated and Arranged by Himself. Edited, with a Biographical Sketch and a brief History of Blindfold Chess, by P. ANDERSON GRAHAM. With Portrait of Mr. Blackburne. 8vo, 7s. 6d. net.

Dead Shot (The): or, Sportsman's Complete Guide. Being a Treatise on the use of the Gun, with Rudimentary and Finishing Lessons in the Art of Shooting Game of all kinds. Also Game-driving, Wildfowl and Pigeon-Shooting, Dog-breaking, etc. By MARKSMAN. With numerous Illustrations. Crown 8vo, 10s. 6d.

Sport and Pastime—*continued.*

Ellis.—CHESS SPARKS; or, Short and Bright Games of Chess. Collected and Arranged by J. H. ELLIS, M.A. 8vo, 4s. 6d.

Folkard.—THE WILD-FOWLER: A Treatise on Fowling, Ancient and Modern, descriptive also of Decoys and Flight-ponds, Wild-fowl Shooting, Gunning-punts, Shooting-yachts, etc. Also Fowling in the Fens and in Foreign Countries, Rock-fowling, etc., etc. By H. C. FOLKARD. With 13 Engravings on Steel, and several Woodcuts. 8vo, 12s. 6d.

Ford.—THE THEORY AND PRACTICE OF ARCHERY. By HORACE FORD. New Edition, thoroughly Revised and Rewritten by W. BUTT, M.A. With a Preface by C. J. LONGMAN, M.A. 8vo, 14s.

Francis.—A BOOK ON ANGLING: or, Treatise on the Art of Fishing in every Branch; including full illustrated List of Salmon Flies. By FRANCIS FRANCIS. With Portrait and Coloured Plates. Crown 8vo, 15s.

Fremantle.—THE BOOK OF THE RIFLE. By the Hon. T. F. FREMANTLE, V.D., Major, 1st Bucks V.R.C. With 54 Plates and 107 Diagrams in the Text. 8vo, 12s. 6d. net.

Gathorne-Hardy.—AUTUMNS IN ARGYLESHIRE WITH ROD AND GUN. By the Hon. A. E. GATHORNE-HARDY. With 8 Photogravure Illustrations by ARCHIBALD THORBURN. 8vo, 6s. net.

Graham. — COUNTRY PASTIMES FOR BOYS. By P. ANDERSON GRAHAM. With 252 Illustrations from Drawings and Photographs. Crown 8vo, gilt edges, 3s. net.

Hutchinson.—THE BOOK OF GOLF AND GOLFERS. By HORACE G. HUTCHINSON. With Contributions by Miss AMY PASCOE, H. H. HILTON, J. H. TAYLOR, H. J. WHIGHAM and Messrs. SUTTON & SONS. With 71 Portraits from Photographs. Large Crown 8vo, gilt top, 7s. 6d. net.

Lang.—ANGLING SKETCHES. By ANDREW LANG. With 20 Illustrations. Crown 8vo, 3s. 6d.

Lillie.—CROQUET UP TO DATE. Containing the Ideas and Teachings of the Leading Players and Champions. By ARTHUR LILLIE. With Contributions by Lieut.-Col. the Hon. H. NEEDHAM, C. D. LOCOCK, etc. With 19 Illustrations (15 Portraits) and numerous Diagrams. 8vo, 10s. 6d. net.

Locock.—SIDE AND SCREW: being Notes on the Theory and Practice of the Game of Billiards. By C. D. LOCOCK. With Diagrams. Crown 8vo, 5s. net.

Longman.—CHESS OPENINGS. By FREDERICK W. LONGMAN. Fcp. 8vo, 2s. 6d.

Mackenzie.—NOTES FOR HUNTING MEN. By Captain CORTLANDT GORDON MACKENZIE. Crown 8vo, 2s. 6d. net.

Madden.—THE DIARY OF MASTER WILLIAM SILENCE: a Study of Shakespeare and of Elizabethan Sport. By the Right Hon. D. H. MADDEN, Vice-Chancellor of the University of Dublin. 8vo, gilt top, 16s.

Maskelyne.—SHARPS AND FLATS: a Complete Revelation of the Secrets of Cheating at Games of Chance and Skill. By JOHN NEVIL MASKELYNE, of the Egyptian Hall. With 62 Illustrations. Crown 8vo, 6s.

Millais (JOHN GUILLE).
THE WILD-FOWLER IN SCOTLAND. With a Frontispiece in Photogravure by Sir J. E. MILLAIS, Bart., P.R.A., 8 Photogravure Plates, 2 Coloured Plates, and 50 Illustrations from the Author's Drawings and from Photographs. Royal 4to, gilt top, 30s. net.

THE NATURAL HISTORY OF THE BRITISH SURFACE-FEEDING DUCKS. With 6 Photogravures and 66 Plates (41 in Colours) from Drawings by the Author, ARCHIBALD THORBURN, and from Photographs. Royal 4to, cloth, gilt top, £6 6s. net.

Modern Bridge. By 'Slam'. With a Reprint of the Laws of Bridge, as adopted by the Portland and Turf Clubs. 18mo, gilt edges, 3s. 6d. net.

Park.—THE GAME OF GOLF. By WILLIAM PARK, Jun., Champion Golfer, 1887-89. With 17 Plates and 26 Illustrations in the Text. Crown 8vo, 7s. 6d.

Sport and Pastime—*continued.*

Payne-Gallwey (Sir RALPH, Bart.).

THE CROSS-BOW : its History, Construction and Management. With numerous Illustrations. Medium 4to.

LETTERS TO YOUNG SHOOTERS (First Series). On the choice and Use of a Gun. With 41 Illustrations. Crown 8vo, 7s. 6d.

LETTERS TO YOUNG SHOOTERS (Second Series). On the Production, Preservation, and Killing of Game. With Directions in Shooting Wood-Pigeons and Breaking-in Retrievers. With Portrait and 103 Illustrations. Crown 8vo, 12s. 6d.

LETTERS TO YOUNG SHOOTERS (Third Series). Comprising a Short Natural History of the Wildfowl that are Rare or Common to the British Islands, with Complete Directions in Shooting Wildfowl on the Coast and Inland. With 200 Illustrations. Cr. 8vo, 18s.

Pole.—THE THEORY OF THE MODERN SCIENTIFIC GAME OF WHIST. By WILLIAM POLE, F.R.S. Fcp. 8vo, gilt edges, 2s. net.

Proctor.—HOW TO PLAY WHIST : with the Laws and Etiquette of Whist. By RICHARD A. PROCTOR. Crown 8vo, gilt edges, 3s. net.

Ronalds.—THE FLY-FISHER'S ENTOMOLOGY. By ALFRED RONALDS. With 20 Coloured Plates. 8vo, 14s.

Selous. — SPORT AND TRAVEL, EAST AND WEST. By FREDERICK COURTENEY SELOUS. With 18 Plates and 35 Illustrations in the Text. Medium 8vo, 12s. 6d. net.

Mental, Moral and Political Philosophy.

LOGIC, RHETORIC, PSYCHOLOGY, ETC.

Abbott.—THE ELEMENTS OF LOGIC. By T. K. ABBOTT, B.D. 12mo, 3s.

Aristotle.

THE ETHICS: Greek Text, Illustrated with Essay and Notes. By Sir ALEXANDER GRANT, Bart. 2 vols. 8vo, 32s.

AN INTRODUCTION TO ARISTOTLE'S ETHICS. Books I.-IV. (Book X., c. vi.-ix. in an Appendix.) With a continuous Analysis and Notes. By the Rev. E. MOORE, D.D. Crown 8vo, 10s. 6d.

Bacon (FRANCIS).

COMPLETE WORKS. Edited by R. L. ELLIS, JAMES SPEDDING and D. D. HEATH. 7 vols. 8vo, £3 13s. 6d.

Bacon (FRANCIS)—*continued.*

LETTERS AND LIFE, including all his occasional Works. Edited by JAMES SPEDDING. 7 vols. 8vo, £4 4s.

THE ESSAYS : With Annotations. By RICHARD WHATELY, D.D. 8vo, 10s. 6d.

THE ESSAYS : With Notes by F. STORR and C. H. GIBSON. Crown 8vo, 3s. 6d.

THE ESSAYS : With Introduction, Notes and Index. By E. A. ABBOTT, D.D. 2 vols. Fcp. 8vo, 6s. The Text and Index only, without Introduction and Notes, in one volume Fcp. 8vo, 2s. 6d.

Mental, Moral and Political Philosophy—*continued.*

Bain (ALEXANDER).

MENTAL AND MORAL SCIENCE: a Compendium of Psychology and Ethics. Crown 8vo, 10s. 6d.

Or Separately,

Part I. PSYCHOLOGY AND HIS-TORY OF PHILOSOPHY. Crown 8vo, 6s. 6d.

Part II. THEORY OF ETHICS AND ETHICAL SYSTEMS. Cr. 8vo, 4s. 6d.

LOGIC. Part I. DEDUCTION. Crown 8vo, 4s. Part II. INDUCTION. Crown 8vo, 6s. 6d.

THE SENSES AND THE INTELLECT. 8vo, 15s.

THE EMOTIONS AND THE WILL. 8vo, 15s.

PRACTICAL ESSAYS. Cr. 8vo, 2s.

Brooks.—THE ELEMENTS OF MIND: being an Examination into the Nature of the First Division of the Elementary Substances of Life. By H. JAMYN BROOKS. 8vo, 10s. 6d. net.

Crozier (JOHN BEATTIE).

CIVILIZATION AND PROGRESS: being the Outlines of a New System of Political, Religious and Social Philosophy. 8vo, 14s.

HISTORY OF INTELLECTUAL DE-VELOPMENT: on the Lines of Modern Evolution.

Vol. I. 8vo, 14s.
Vol. II. (*In preparation.*)
Vol. III. 8vo, 10s. 6d.

Davidson.—THE LOGIC OF DE-FINITION, Explained and Applied. By WILLIAM L. DAVIDSON, M.A. Cr. 8vo, 6s.

Green (THOMAS HILL).—THE WORKS OF. Edited by R. L. NETTLESHIP.

Vols. I. and II. Philosophical Works. 8vo, 16s. each.

Vol. III. Miscellanies. With Index to the three Volumes, and Memoir. 8vo, 21s.

LECTURES ON THE PRINCIPLES OF POLITICAL OBLIGATION. With Preface by BERNARD BOSAN-QUET. 8vo, 5s.

Gurnhill.—THE MORALS OF SUI-CIDE. By the Rev. J. GURNHILL, B.A. Vol. I., Cr. 8vo, 5s. net. Vol. II., Cr. 8vo, 5s. net.

Hodgson (SHADWORTH H.).

TIME AND SPACE: a Metaphysical Essay. 8vo, 16s.

THE THEORY OF PRACTICE: an Ethical Inquiry. 2 vols. 8vo, 24s.

THE PHILOSOPHY OF REFLEC-TION. 2 vols. 8vo, 21s.

THE METAPHYSIC OF EXPERI-ENCE. Book I. General Analysis of Experience; Book II. Positive Science; Book III. Analysis of Conscious Action; Book IV. The Real Universe. 4 vols. 8vo, 36s. net.

Hume.—THE PHILOSOPHICAL WORKS OF DAVID HUME. Edited by T. H. GREEN and T. H. GROSE. 4 vols. 8vo, 28s. Or separately. Essays. 2 vols. 14s. Treatise of Human Nature. 2 vols. 14s.

James (WILLIAM, M.D., LL.D.).

THE WILL TO BELIEVE, and Other Essays in Popular Philosophy. Crown 8vo, 7s. 6d.

THE VARIETIES OF RELIGIOUS EXPERIENCE: a Study in Human Nature. Being the Gifford Lectures on Natural Religion delivered at Edinburgh in 1901-1902. 8vo, 12s. net.

Justinian.—THE INSTITUTES OF JUSTINIAN: Latin Text, chiefly that of Huschke, with English Introduction, Translation, Notes and Summary. By THOMAS C. SANDARS, M.A. 8vo, 18s.

Kant (IMMANUEL).

CRITIQUE OF PRACTICAL REASON, AND OTHER WORKS ON THE THEORY OF ETHICS. Translated by T. K. ABBOTT, B.D. With Memoir. 8vo, 12s. 6d.

FUNDAMENTAL PRINCIPLES OF THE METAPHYSIC OF ETHICS. Translated by T. K. ABBOTT, B.D. Crown 8vo, 3s.

INTRODUCTION TO LOGIC, AND HIS ESSAY ON THE MISTAKEN SUBTILTY OF THE FOUR FIGURES. Translated by T. K. ABBOTT. 8vo, 6s.

Mental, Moral and Political Philosophy—*continued.*

Kelly.—GOVERNMENT OR HUMAN EVOLUTION. By EDMOND KELLY, M.A., F.G.S. Vol. I. Justice. Crown 8vo, 7s. 6d. net. Vol. II. Collectivism and Individualism. Cr. 8vo, 10s. 6d. net.

Killick.—HANDBOOK TO MILL'S SYSTEM OF LOGIC. By Rev. A. H. KILLICK, M.A. Crown 8vo, 3s. 6d.

Ladd (GEORGE TRUMBULL).

PHILOSOPHY OF CONDUCT : a Treatise of the Facts, Principles and Ideals of Ethics. 8vo, 21s.

ELEMENTS OF PHYSIOLOGICAL PSYCHOLOGY. 8vo, 21s.

OUTLINES OF DESCRIPTIVE PSYCHOLOGY : a Text-Book of Mental Science for Colleges and Normal Schools. 8vo, 12s.

OUTLINES OF PHYSIOLOGICAL PSYCHOLOGY. 8vo, 12s.

PRIMER OF PSYCHOLOGY. Crown 8vo, 5s. 6d.

Lecky.—THE MAP OF LIFE : Conduct and Character. By WILLIAM EDWARD HARTPOLE LECKY. Library Edition, 8vo, 10s. 6d. Cabinet Edition, Crown 8vo, 5s. net.

Lutoslawski.—THE ORIGIN AND GROWTH OF PLATO'S LOGIC. With an Account of Plato's Style and of the Chronology of his Writings. By WINCENTY LUTOSLAWSKI. 8vo, 21s.

Max Müller (F.).

THE SCIENCE OF THOUGHT. 8vo, 21s.

THE SIX SYSTEMS OF INDIAN PHILOSOPHY. 8vo, 18s.

THREE LECTURES ON THE VEDANTA PHILOSOPHY. Cr. 8vo, 5s.

Mill (JOHN STUART).

A SYSTEM OF LOGIC. Cr. 8vo, 3s. 6d.

ON LIBERTY. Crown 8vo, 1s. 4d.

CONSIDERATIONS ON REPRESENTATIVE GOVERNMENT. Crown 8vo, 2s.

Mill (JOHN STUART)—*continued.*

UTILITARIANISM. 8vo, 2s. 6d.

EXAMINATION OF SIR WILLIAM HAMILTON'S PHILOSOPHY. 8vo, 16s.

NATURE, THE UTILITY OF RELIGION AND THEISM. Three Essays. 8vo, 5s.

Monck.—AN INTRODUCTION TO LOGIC. By WILLIAM HENRY S. MONCK, M.A. Crown 8vo, 5s.

Pierce.—STUDIES IN AUDITORY AND VISUAL SPACE PERCEPTION : Essays on Experimental Psychology. By A. H. PIERCE. Cr. 8vo, 6s. 6d. net.

Richmond. — THE MIND OF A CHILD. By ENNIS RICHMOND. Crown 8vo, 3s. 6d. net.

Romanes.—MIND AND MOTION AND MONISM. By GEORGE JOHN ROMANES. Crown 8vo, 4s. 6d.

Sully (JAMES).

AN ESSAY ON LAUGHTER : its Forms, its Cause, its Development and its Value. 8vo, 12s. 6d. net.

THE HUMAN MIND : a Text-book of Psychology. 2 vols. 8vo, 21s.

OUTLINES OF PSYCHOLOGY. Cr. 8vo, 9s.

THE TEACHER'S HANDBOOK OF PSYCHOLOGY. Crown 8vo, 6s. 6d.

STUDIES OF CHILDHOOD. 8vo, 10s. 6d.

CHILDREN'S WAYS : being Selections from the Author's 'Studies of Childhood'. With 25 Illustrations. Crown 8vo, 4s. 6d.

Sutherland.—THE ORIGIN AND GROWTH OF THE MORAL INSTINCT. By ALEXANDER SUTHERLAND, M.A. 2 vols. 8vo, 28s.

Swinburne.—PICTURE LOGIC : an Attempt to Popularise the Science of Reasoning. By ALFRED JAMES SWINBURNE, M.A. With 23 Woodcuts. Crown 8vo, 2s. 6d.

Mental, Moral and Political Philosophy—*continued*.

Thomas. — INTUITIVE SUGGES-TION. By J. W. THOMAS, Author of 'Spiritual Law in the Natural World,' etc. Crown 8vo, 3s. 6d. net.

Webb.—THE VEIL OF ISIS ; a Series of Essays on Idealism. By THOMAS E. WEBB, LL.D., Q.C. 8vo, 10s. 6d.

Weber.—HISTORY OF PHILOSO-PHY. By ALFRED WEBER, Professor in the University of Strasburg. Translated by FRANK THILLY, Ph.D. 8vo, 16s.

Whately (ARCHBISHOP).
BACON'S ESSAYS. With Annotations. 8vo, 10s. 6d.
ELEMENTS OF LOGIC. Crown 8vo, 4s. 6d.
ELEMENTS OF RHETORIC. Crown 8vo, 4s. 6d.

Zeller (Dr. EDWARD).

THE STOICS, EPICUREANS, AND SCEPTICS. Translated by the Rev. O. J. REICHEL, M.A. Crown 8vo, 15s.

OUTLINES OF THE HISTORY OF GREEK PHILOSOPHY. Translated by SARAH F. ALLEYNE and EVELYN ABBOTT, M.A., LL.D. Cr. 8vo, 10s. 6d.

PLATO AND THE OLDER ACA-DEMY. Translated by SARAH F. ALLEYNE and ALFRED GOODWIN, B.A. Crown 8vo, 18s.

SOCRATES AND THE SOCRATIC SCHOOLS. Translated by the Rev. O. J. REICHEL, M.A. Cr. 8vo, 10s. 6d.

ARISTOTLE AND THE EARLIER PERIPATETICS. Translated by B. F. C. COSTELLOE, M.A., and J. H. MUIRHEAD, M.A. 2 vols. Cr. 8vo, 24s.

STONYHURST PHILOSOPHICAL SERIES.

A MANUAL OF POLITICAL ECO-NOMY. By C. S. DEVAS, M.A. Crown 8vo, 7s. 6d.

FIRST PRINCIPLES OF KNOW-LEDGE. By JOHN RICKABY, S.J. Crown 8vo, 5s.

GENERAL METAPHYSICS. By JOHN RICKABY, S.J. Crown 8vo, 5s.

LOGIC. By RICHARD F. CLARKE, S.J. Crown 8vo, 5s.

MORAL PHILOSOPHY (ETHICS AND NATURAL LAW). By JOSEPH RICKABY, S.J. Crown 8vo, 5s.

NATURAL THEOLOGY. By BERNARD BOEDDER, S.J. Crown 8vo, 6s. 6d.

PSYCHOLOGY. By MICHAEL MAHER, S.J., D.Litt., M.A. (Lond.). Crown 8vo, 6s. 6d.

History and Science of Language, etc.

Davidson.—LEADING AND IM-PORTANT ENGLISH WORDS : Ex-plained and Exemplified. By WILLIAM L. DAVIDSON, M.A. Fcp. 8vo, 3s. 6d.

Farrar.—LANGUAGE AND LAN-GUAGES. By F. W. FARRAR, D.D., Dean of Canterbury. Crown 8vo, 6s.

Graham. — ENGLISH SYNONYMS, Classified and Explained : with Practical Exercises. By G. F. GRAHAM. Fcp. 8vo, 6s.

Max Müller (F.).

THE SCIENCE OF LANGUAGE. 2 vols. Crown 8vo, 10s.

Max Müller (F.)--*continued*.
BIOGRAPHIES OF WORDS, AND THE HOME OF THE ARYAS. Crown 8vo, 5s.
CHIPS FROM A GERMAN WORK-SHOP. Vol. III. ESSAYS ON LANGUAGE AND LITERATURE. Crown 8vo, 5s.
LAST ESSAYS. First Series. Essays on Language, Folklore and other Subjects. Crown 8vo, 5s.

Roget. - THESAURUS OF ENGLISH WORDS AND PHRASES. Classified and Arranged so as to Facilitate the Expression of Ideas and Assist in Lite-rary Composition. By PETER MARK ROGET. M.D., F.R.S. With full Index. Cr. 8vo, 9s. net.

Political Economy and Economics.

Ashley (W. J.).
ENGLISH ECONOMIC HISTORY AND THEORY. Crown 8vo, Part I., 5s. Part II., 10s. 6d.
SURVEYS, HISTORIC AND ECONOMIC. Crown 8vo, 9s. net.

Bagehot.—ECONOMIC STUDIES. By WALTER BAGEHOT. Crown 8vo, 3s. 6d.

Barnett. — PRACTICABLE SOCIALISM. Essays on Social Reform. By SAMUEL A. and HENRIETTA BARNETT. Crown 8vo, 6s.

Devas.—A MANUAL OF POLITICAL ECONOMY. By C. S. DEVAS, M.A. Crown 8vo, 7s. 6d. (*Stonyhurst Philosophical Series.*)

Lawrence. — LOCAL VARIATIONS IN WAGES. By F. W. LAWRENCE, M.A. With Index and 18 Maps and Diagrams. 4to, 8s. 6d.

Leslie. — ESSAYS ON POLITICAL ECONOMY. By T. E. CLIFFE LESLIE, Hon. LL.D., Dubl. 8vo, 10s. 6d.

Macleod (HENRY DUNNING).
BIMETALLISM. 8vo, 5s. net.
THE ELEMENTS OF BANKING. Cr. 8vo, 3s. 6d.
THE THEORY AND PRACTICE OF BANKING. Vol. I. 8vo, 12s. Vol. II. 14s.

Macleod (HENRY DUNNING)—*cont.*
THE THEORY OF CREDIT. 8vo. In 1 vol. 30s. net; or separately, Vol. I., 10s. net. Vol. II., Part I., 10s. net. Vol. II., Part II., 10s. net.
INDIAN CURRENCY. 8vo, 2s. 6d. net.

Mill.—POLITICAL ECONOMY. By JOHN STUART MILL.
Popular Edition. Crown 8vo, 3s. 6d.
Library Edition. 2 vols. 8vo, 30s.

Mulhall. — INDUSTRIES AND WEALTH OF NATIONS. By MICHAEL G. MULHALL, F.S.S. With 32 Diagrams. Crown 8vo, 8s. 6d.

Symes.—POLITICAL ECONOMY: a Short Textbook of Political Economy. With Problems for solution, Hints for Supplementary Reading, and a Supplementary chapter on Socialism. By J. E. SYMES, M.A. Crown 8vo, 2s. 6d.

Toynbee.—LECTURES ON THE INDUSTRIAL REVOLUTION OF THE 18TH CENTURY IN ENGLAND. By ARNOLD TOYNBEE. 8vo, 10s. 6d.

Webb (SIDNEY and BEATRICE).
THE HISTORY OF TRADE UNIONISM. With Map and Bibliography. 8vo, 7s. 6d. net.
INDUSTRIAL DEMOCRACY: a Study in Trade Unionism. 2 vols. 8vo, 12s. net.
PROBLEMS OF MODERN INDUSTRY. 8vo, 5s net.

Evolution, Anthropology, etc.

Avebury. — THE ORIGIN OF CIVILISATION, and the Primitive Condition of Man. By the Right Hon. LORD AVEBURY. With 6 Plates and 20 Illustrations. 8vo, 18s.

Clodd (EDWARD).
THE STORY OF CREATION: a Plain Account of Evolution. With 77 Illustrations. Crown 8vo, 3s. 6d.
A PRIMER OF EVOLUTION: being a Popular Abridged Edition of 'The Story of Creation'. With Illustrations. Fcp. 8vo, 1s. 6d.

Packard.—LAMARCK, THE FOUNDER OF EVOLUTION: his Life and Work, with Translations of his Writings on Organic Evolution. By ALPHEUS S. PACKARD, M.D., LL.D., Professor of Zoology and Geology in Brown University. With 10 Portrait and other Illustrations. Large Crown 8vo, 9s. net.

Romanes (GEORGE JOHN).
ESSAYS. Edited by C. LLOYD MORGAN. Crown 8vo, 5s. net.
AN EXAMINATION OF WEISMANNISM. Crown 8vo, 6s.

Evolution, Anthropology, etc.—*continued*.

Romanes (GEORGE JOHN)—*continued*.
DARWIN, AND AFTER DARWIN:
an Exposition of the Darwinian
Theory, and a Discussion on Post-
Darwinian Questions.
Part I. THE DARWINIAN THEORY.
With Portrait of Darwin and 125
Illustratious. Crown 8vo, 10s. 6d.

Part II. POST-DARWINIAN QUES-
TIONS : Heredity and Utility. With
Portrait of the Author and 5 Illus-
trations. Crown 8vo, 10s. 6d.

Part III. POST-DARWINIAN QUES-
TIONS : Isolation and Physiological
Selection. Crown 8vo, 5s.

The Science of Religion, etc.

Balfour.—THE FOUNDATIONS OF
BELIEF : being Notes Introductory to
the Study of Theology. By the Right
Hon. ARTHUR JAMES BALFOUR. Crown
8vo, 6s. net.

Baring-Gould.—THE ORIGIN AND
DEVELOPMENT OF RELIGIOUS
BELIEF. By the Rev. S. BARING-
GOULD. 2 vols. Crown 8vo, 3s. 6d. each.

Campbell.—RELIGION IN GREEK
LITERATURE. By the Rev. LEWIS
CAMPBELL, M.A., LL.D. 8vo, 15s.

Davidson.—THEISM, as Grounded in
Human Nature, Historically and Critic-
ally Handled. Being the Burnett
Lectures for 1892 and 1893, delivered at
Aberdeen. By W. L. DAVIDSON, M.A.,
LL.D. 8vo, 15s.

James.—THE VARIETIES OF RE-
LIGIOUS EXPERIENCE : a Study in
Human Nature. Being the Gifford
Lectures on Natural Religion delivered
at Edinburgh in 1901-1902. By WILLIAM
JAMES, LL.D., etc. 8vo, 12s. net.

Lang (ANDREW).
MAGIC AND RELIGION. 8vo, 10s. 6d.
CUSTOM AND MYTH : Studies of
Early Usage and Belief. With 15
Illustrations. Crown 8vo, 3s. 6d.
MYTH, RITUAL AND RELIGION.
2 vols. Crown 8vo, 7s.
MODERN MYTHOLOGY : a Reply to
Professor Max Müller. 8vo, 9s.
THE MAKING OF RELIGION. Cr.
8vo, 5s. net.

Leighton. — TYPICAL MODERN
CONCEPTIONS OF GOD ; or, The
Absolute of German Romantic Idealism
and of English Evolutionary Agnos-
ticism. By JOSEPH ALEXANDER
LEIGHTON, Professor of Philosophy in
Hobart College, U.S. Crown 8vo, 3s. 6d.
net.

Max Müller (The Right Hon. F.).

CHIPS FROM A GERMAN WORK-
SHOP. Vol. IV. Essays on Mythology and Folk Lore. Crown 8vo, 5s.

THE SIX SYSTEMS OF INDIAN
PHILOSOPHY. 8vo, 18s.

CONTRIBUTIONS TO THE SCIENCE
OF MYTHOLOGY. 2 vols. 8vo, 32s.

THE ORIGIN AND GROWTH OF RE-
LIGION, as illustrated by the Religions
of India. The Hibbert Lectures, delivered at the Chapter House, Westminster Abbey, in 1878. Cr. 8vo, 5s.

INTRODUCTION TO THE SCIENCE
OF RELIGION : Four Lectures de-
livered at the Royal Institution.
Crown 8vo, 5s.

NATURAL RELIGION. The Gifford
Lectures, delivered before the University of Glasgow in 1888. Cr. 8vo, 5s.

PHYSICAL RELIGION. The Gifford
Lectures, delivered before the University of Glasgow in 1890. Cr. 8vo, 5s.

ANTHROPOLOGICAL RELIGION.
The Gifford Lectures, delivered before
the University of Glasgow in 1891.
Crown 8vo, 5s.

THEOSOPHY ; or, PSYCHOLOGICAL
RELIGION. The Gifford Lectures,
delivered before the University of
Glasgow in 1892. Crown 8vo, 5s.

THREE LECTURES ON THE
VEDANTA PHILOSOPHY, de-
livered at the Royal Institution in
March, 1894. Crown 8vo, 5s.

LAST ESSAYS. Second Series—Essays
on the Science of Religion. Cr. 8vo, 5s.

The Science of Religion, etc.—*continued.*

Oakesmith.—THE RELIGION OF PLUTARCH: a Pagan Creed of Apostolic Times. An Essay. By JOHN OAKESMITH, D. Litt., M.A. Crown 8vo, 5s. net.

Wood-Martin (W. G.).
TRACES OF THE ELDER FAITHS OF IRELAND: a Folklore Sketch. A Handbook of Irish Pre-Christian Traditions. With 192 Illustrations. 2 vols. 8vo, 30s. net.
PAGAN IRELAND: an Archæological Sketch. A Handbook of Irish Pre-Christian Antiquities. With 512 Illustrations. 8vo, 15s.

Classical Literature, Translations, etc.

Abbott.—HELLENICA. A Collection of Essays on Greek Poetry, Philosophy, History and Religion. Edited by EVELYN ABBOTT, M.A., LL.D. Crown 8vo, 7s. 6d.

Æschylus. — EUMENIDES OF ÆSCHYLUS. With Metrical English Translation. By J. F. DAVIES. 8vo, 7s.

Aristophanes.—THE ACHARNIANS OF ARISTOPHANES, translated into English Verse. By R. Y. TYRRELL. Crown 8vo, 1s.

Becker (W. A.). Translated by the Rev. F. METCALFE, B.D.
GALLUS: or, Roman Scenes in the Time of Augustus. With Notes and Excursuses. With 26 Illustrations. Crown 8vo, 3s. 6d.
CHARICLES: or, Illustrations of the Private Life of the Ancient Greeks. With Notes and Excursuses. With 26 Illustrations. Crown 8vo, 3s. 6d.

Campbell.—RELIGION IN GREEK LITERATURE. By the Rev. LEWIS CAMPBELL, M.A., LL.D., Emeritus Professor of Greek, University of St. Andrews. 8vo, 15s.

Cicero. — CICERO'S CORRESPONDENCE. By R. Y. TYRRELL. Vols. I., II., III., 8vo, each 12s. Vol. IV., 15s. Vol. V., 14s. Vol. VI., 12s. Vol. VII., Index, 7s. 6d.

Harvard Studies in Classical Philology. Edited by a Committee of the Classical Instructors of Harvard University. Vols. XI., 1900; XII., 1901; XIII., 1902. 8vo, 6s. 6d. net each.

Hime. — LUCIAN, THE SYRIAN SATIRIST. By Lieut.-Colonel HENRY W. L. HIME (late) Royal Artillery. 8vo, 5s. net.

Homer.—THE ODYSSEY OF HOMER. Done into English Verse. By WILLIAM MORRIS. Crown 8vo, 6s.

Horace.—THE WORKS OF HORACE, rendered into English Prose. With Life, Introduction and Notes. By WILLIAM COUTTS, M.A. Crown 8vo, 5s. net.

Lang.—HOMER AND THE EPIC. By ANDREW LANG. Crown 8vo, 9s. net.

Lucian.—TRANSLATIONS FROM LUCIAN. By AUGUSTA M. CAMPBELL DAVIDSON, M.A. Edin. Crown 8vo, 5s. net.

Ogilvie.—HORAE LATINAE: Studies in Synonyms and Syntax. By the late ROBERT OGILVIE, M.A., LL.D., H.M. Chief Inspector of Schools for Scotland. Edited by ALEXANDER SOUTER, M.A. With a Memoir by JOSEPH OGILVIE, M.A., LL.D. 8vo, 12s. 6d. net.

Rich.—A DICTIONARY OF ROMAN AND GREEK ANTIQUITIES. By A. RICH, B.A. With 2000 Woodcuts. Crown 8vo, 6s. net.

Sophocles.—Translated into English Verse. By ROBERT WHITELAW, M.A., Assistant Master in Rugby School. Cr. 8vo, 8s. 6d.

Tyrrell.—DUBLIN TRANSLATIONS INTO GREEK AND LATIN VERSE. Edited by R. Y. TYRRELL. 8vo, 6s.

Classical Literature, Translations, etc.—*continued.*

Virgil.

THE POEMS OF VIRGIL. Translated into English Prose by JOHN CONINGTON. Crown 8vo, 6s.

THE ÆNEID OF VIRGIL. Translated into English Verse by JOHN CONINGTON. Crown 8vo, 6s.

THE ÆNEIDS OF VIRGIL. Done into English Verse. By WILLIAM MORRIS. Crown 8vo, 6s.

THE ÆNEID OF VIRGIL, freely translated into English Blank Verse. By W. J. THORNHILL. Crown 8vo, 6s. net.

Virgil—*continued.*

THE ÆNEID OF VIRGIL. Translated into English Verse by JAMES RHOADES.
Books I.-VI. Crown 8vo, 5s.
Books VII.-XII. Crown 8vo, 5s.

THE ECLOGUES AND GEORGICS OF VIRGIL. Translated into English Prose by J. W. MACKAIL, Fellow of Balliol College, Oxford. 16mo, 5s.

Wilkins.—THE GROWTH OF THE HOMERIC POEMS. By G. WILKINS. 8vo, 6s.

Poetry and the Drama.

Arnold.—THE LIGHT OF THE WORLD; or, the Great Consummation. By Sir EDWIN ARNOLD. With 14 Illustrations after HOLMAN HUNT. Crown 8vo, 5s. net.

Bell (Mrs. HUGH).
CHAMBER COMEDIES: a Collection of Plays and Monologues for the Drawing-room. Crown 8vo, 5s. net.
FAIRY TALE PLAYS, AND HOW TO ACT THEM. With 91 Diagrams and 52 Illustrations. Crown 8vo, 3s. net.
RUMPELSTILTZKIN: a Fairy Play in Five Scenes (Characters, 7 Male; 1 Female). From 'Fairy Tale Plays and How to Act Them'. With Illustrations, Diagrams and Music. Crown 8vo, sewed, 6d.

Bird.—RONALD'S FAREWELL, and other Verses. By GEORGE BIRD, M.A., Vicar of Bradwell, Derbyshire. Fcp. 8vo, 4s. 6d. net.

Dabney.—THE MUSICAL BASIS OF VERSE: a Scientific Study of the Principles of Poetic Composition. By J. P. DABNEY. Crown 8vo, 6s.6d. net.

Ingelow (JEAN).
POETICAL WORKS. Complete in One Volume. Crown 8vo, gilt top, 6s. net.
LYRICAL AND OTHER POEMS. Selected from the Writings of JEAN INGELOW. Fcp. 8vo, 2s. 6d. cloth plain, 3s. cloth gilt.

Keary.—THE BROTHERS: a Fairy Masque. By C. F. KEARY. Crown 8vo, 4s. net.

Lang (ANDREW).
GRASS OF PARNASSUS. Fcp. 8vo, 2s. 6d. net.
THE BLUE POETRY BOOK. Edited by ANDREW LANG. With 100 Illustrations. Crown 8vo, gilt edges, 6s.

Lecky.—POEMS. By the Right Hon. W. E. H. LECKY. Fcp. 8vo, 5s.

Lytton (THE EARL OF), (OWEN MEREDITH).
THE WANDERER. Cr. 8vo, 10s. 6d.
LUCILE. Crown 8vo, 10s. 6d.
SELECTED POEMS. Cr. 8vo, 10s. 6d.

Macaulay.—LAYS OF ANCIENT ROME, WITH 'IVRY' AND 'THE ARMADA'. By Lord MACAULAY. Illustrated by G. SCHARF. Fcp. 4to, 10s. 6d.
———— Bijou Edition, 18mo, 2s. 6d., gilt top.
———— Popular Edition, Fcp. 4to, 6d. sewed, 1s. cloth. Illustrated by J. R. WEGUELIN. Cr. 8vo, 3s. net.
Annotated Edition. Fcp. 8vo, 1s. sewed, 1s. 6d. cloth.

Poetry and the Drama—*continued*.

MacDonald.—A BOOK OF STRIFE, IN THE FORM OF THE DIARY OF AN OLD SOUL: Poems. By GEORGE MACDONALD, LL.D. 18mo, 6s.

Morris (WILLIAM).

POETICAL WORKS—LIBRARY EDITION.

Complete in 11 volumes. Crown 8vo, price 5s. net each.

THE EARTHLY PARADISE. 4 vols. Crown 8vo, 5s. net each.

THE LIFE AND DEATH OF JASON. Crown 8vo, 5s. net.

THE DEFENCE OF GUENEVERE, and other Poems. Crown 8vo, 5s. net.

THE STORY OF SIGURD THE VOLSUNG, AND THE FALL OF THE NIBLUNGS. Crown 8vo, 5s. net.

POEMS BY THE WAY, AND LOVE IS ENOUGH. Crown 8vo, 5s. net.

THE ODYSSEY OF HOMER. Done into English Verse. Crown 8vo, 5s. net.

THE ÆNEIDS OF VIRGIL. Done into English Verse. Crown 8vo, 5s. net.

THE TALE OF BEOWULF, SOMETIME KING OF THE FOLK OF THE WEDERGEATS. Translated by WILLIAM MORRIS and A. J. WYATT. Crown 8vo, 5s. net.

Certain of the POETICAL WORKS may also be had in the following Editions :—

THE EARTHLY PARADISE.

Popular Edition. 5 Vols. 12mo, 25s. ; or 5s. each, sold separately.

The same in Ten Parts, 25s. ; or 2s. 6d. each, sold separately.

Cheap Edition, in 1 vol. Crown 8vo, 6s. net.

POEMS BY THE WAY. Square crown 8vo, 6s.

*** For Mr. William Morris's other Works, see pp. 27, 37, 40.

Morte Arthur: an Alliterative Poem of the Fourteenth Century. Edited from the Thornton MS., with Introduction, Notes and Glossary. By MARY MACLEOD BANKS. Fcp. 8vo, 3s. 6d.

Nesbit.—LAYS AND LEGENDS. By E. NESBIT (Mrs. HUBERT BLAND). First Series. Crown 8vo, 3s. 6d. Second Series. With Portrait. Crown 8vo, 5s.

Ramal.—SONGS OF CHILDHOOD. By WALTER RAMAL. With a Frontispiece from a Drawing by RICHARD DOYLE. Fcp. 8vo, 3s. 6d. net.

Riley. — OLD-FASHIONED ROSES: Poems. By JAMES WHITCOMBE RILEY. 12mo, gilt top, 5s.

Romanes.—A SELECTION FROM THE POEMS OF GEORGE JOHN ROMANES, M.A., LL.D., F.R.S. With an Introduction by T. HERBERT WARREN, President of Magdalen College, Oxford. Crown 8vo, 4s. 6d.

Savage-Armstrong.--BALLADS OF DOWN. By G. F. SAVAGE-ARMSTRONG. M.A., D.Litt. Crown 8vo, 7s. 6d.

Shakespeare.

BOWDLER'S FAMILY SHAKESPEARE. With 36 Woodcuts. 1 vol. 8vo, 14s. Or in 6 vols. Fcp. 8vo, 21s.

THE SHAKESPEARE BIRTHDAY BOOK. By MARY F. DUNBAR. 32mo, 1s. 6d.

Stevenson.—A CHILD'S GARDEN OF VERSES. By ROBERT LOUIS STEVENSON. Fcp. 8vo, gilt top, 5s.

Wagner. — THE NIBELUNGEN RING. Done into English Verse by REGINALD RANKIN, B.A., of the Inner Temple, Barrister-at-Law.

Vol. I. Rhine Gold, The Valkyrie. Fcp. 8vo, gilt top, 4s. 6d.

Vol. II. Siegfried, The Twilight of the Gods. Fcp. 8vo, gilt top, 4s. 6d.

Fiction, Humour, etc.

Anstey (F.).

VOCES POPULI. (Reprinted from *Punch*.)

First Series. With 20 Illustrations by J. BERNARD PARTRIDGE. Crown 8vo, gilt top, 3s. net.

Second Series. With 25 Illustrations by J. BERNARD PARTRIDGE. Crown 8vo, gilt top, 3s. net.

THE MAN FROM BLANKLEY'S, and other Sketches. (Reprinted from *Punch*.) With 25 Illustrations by J. BERNARD PARTRIDGE. Crown 8vo, gilt top, 3s. net.

Bailey.—MY LADY OF ORANGE: a Romance of the Netherlands in the Days of Alva. By H. C. BAILEY. With 8 Illustrations. Crown 8vo, 6s.

Beaconsfield (THE EARL OF).

NOVELS AND TALES. Complete in 11 vols. Crown 8vo, 1s. 6d. each, or in sets, 11 vols., gilt top, 15s. net.

Vivian Grey.	Sybil.
The Young Duke, etc.	Henrietta Temple.
Alroy, Ixion, etc.	Venetia.
Contarini, Fleming, etc.	Coningsby.
	Lothair.
Tancred.	Endymion.

NOVELS AND TALES. THE HUGH-ENDEN EDITION. With 2 Portraits and 11 Vignettes. 11 vols. Crown 8vo, 42s.

Bottome. — LIFE. THE INTER-PRETER. By PHYLLIS BOTTOME. Crown 8vo, 6s.

Churchill.—SAVROLA : a Tale of the Revolution in Laurania. By WINSTON SPENCER CHURCHILL, M.P. Crown 8vo, 6s.

Crawford.—THE AUTOBIOGRAPHY OF A TRAMP. By J. H. CRAWFORD. With a Photogravure Frontispiece 'The Vagrants,' by FRED. WALKER, and 8 other Illustrations. Crown 8vo, 5s. net.

Creed.—THE VICAR OF ST. LUKE'S. By SIBYL CREED. Cr. 8vo, 6s.

Davenport.—BY THE RAMPARTS OF JEZREEL : a Romance of Jehu, King of Israel. By ARNOLD DAVENPORT. With Frontispiece by LANCELOT SPEED. Crown 8vo, 6s.

Dougall.—BEGGARS ALL. By L. DOUGALL. Crown 8vo, 3s. 6d.

Doyle (Sir A. CONAN).

MICAH CLARKE : a Tale of Monmouth's Rebellion. With 10 Illustrations. Crown 8vo, 3s. 6d.

THE REFUGEES : a Tale of the Huguenots. With 25 Illustrations. Crown 8vo, 3s. 6d.

THE STARK MUNRO LETTERS. Crown 8vo, 3s. 6d.

THE CAPTAIN OF THE POLESTAR, and other Tales. Crown 8vo, 3s. 6d.

Dyson.—THE GOLD-STEALERS : a Story of Waddy. By EDWARD DYSON, Author of 'Rhymes from the Mines,' etc. Crown 8vo, 6s.

Farrar (F. W., DEAN OF CANTERBURY).

DARKNESS AND DAWN : or, Scenes in the Days of Nero. An Historic Tale. Crown 8vo, gilt top, 6s. net.

GATHERING CLOUDS : a Tale of the Days of St. Chrysostom. Crown 8vo, gilt top, 6s. net.

Fowler (EDITH H.).

THE YOUNG PRETENDERS. A Story of Child Life. With 12 Illustrations by Sir PHILIP BURNE-JONES, Bart. Crown 8vo, 6s.

THE PROFESSOR'S CHILDREN. With 24 Illustrations by ETHEL KATE BURGESS. Crown 8vo, 6s.

Francis (M. E.).

FIANDER'S WIDOW. Crown 8vo, 6s.

YEOMAN FLEETWOOD. With Frontispiece. Crown 8vo, 3s. net.

PASTORALS OF DORSET. With 8 Illustrations. Crown 8vo, 6s.

THE MANOR FARM. With Frontispiece by CLAUD C. DU PRE COOPER. Crown 8vo, 6s.

Fiction, Humour, etc.—*continued.*

Froude.—THE TWO CHIEFS OF DUNBOY : an Irish Romance of the Last Century. By JAMES A. FROUDE. Crown 8vo, 3s. 6d.

Gurdon.—MEMORIES AND FANCIES : Suffolk Tales and other Stories ; Fairy Legends ; Poems ; Miscellaneous Articles. By the late Lady CAMILLA GURDON. Crown 8vo, 5s.

Haggard (H. RIDER).

ALLAN QUATERMAIN. With 31 Illustrations. Crown 8vo, 3s. 6d.

ALLAN'S WIFE. With 34 Illustrations. Crown 8vo, 3s. 6d.

BEATRICE. With Frontispiece and Vignette. Crown 8vo, 3s. 6d.

BLACK HEART AND WHITE HEART, and other Stories. With 33 Illustrations. Crown 8vo, 6s.

CLEOPATRA. With 29 Illustrations. Crown 8vo, 3s. 6d.

COLONEL QUARITCH, V.C. With Frontispiece and Vignette. Crown 8vo, 3s. 6d.

DAWN. With 16 Illustrations. Crown 8vo, 3s. 6d.

DOCTOR THERNE. Cr. 8vo, 3s. 6d.

ERIC BRIGHTEYES. With 51 Illustrations. Crown 8vo, 3s. 6d.

HEART OF THE WORLD. With 15 Illustrations. Crown 8vo, 3s. 6d.

JOAN HASTE. With 20 Illustrations. Crown 8vo, 3s. 6d.

LYSBETH. With 26 Illustrations. Crown 8vo, 6s.

MAIWA'S REVENGE. Cr. 8vo, 1s. 6d.

MONTEZUMA'S DAUGHTER. With 24 Illustrations. Crown 8vo, 3s. 6d.

Haggard (H. RIDER)—*continued.*

MR. MEESON'S WILL. With 16 Illustrations. Crown 8vo, 3s. 6d.

NADA THE LILY. With 23 Illustrations. Crown 8vo, 3s. 6d.

SHE. With 32 Illustrations. Crown 8vo, 3s. 6d.

SWALLOW : a Tale of the Great Trek. With 8 Illustrations. Crown 8vo 3s. 6d.

THE PEOPLE OF THE MIST. With 16 Illustrations. Crown 8vo, 3s. 6d.

THE WITCH'S HEAD. With 16 Illustrations. Crown 8vo, 3s. 6d.

Haggard and Lang. — THE WORLD'S DESIRE. By H. RIDER HAGGARD and ANDREW LANG. With 27 Illustrations. Crown 8vo, 3s. 6d.

Harte. — IN THE CARQUINEZ WOODS. By BRET HARTE. Crown 8vo, 3s. 6d.

Hope.—THE HEART OF PRINCESS OSRA. By ANTHONY HOPE. With 9 Illustrations. Crown 8vo, 3s. 6d.

Howard. — THE FAILURE OF SUCCESS. By Lady MABEL HOWARD. Crown 8vo, 6s.

Hutchinson. — A FRIEND OF NELSON. By HORACE G. HUTCHINSON. Crown 8vo, 6s.

Jerome.—SKETCHES IN LAVENDER : BLUE AND GREEN. By JEROME K. JEROME, Author of 'Three Men in a Boat,' etc. Crown 8vo, 3s. 6d.

Joyce.—OLD CELTIC ROMANCES. Twelve of the most beautiful of the Ancient Irish Romantic Tales. Translated from the Gaelic. By P. W. JOYCE, LL.D. Crown 8vo, 3s. 6d.

Fiction, Humour, etc.—*continued.*

Lang (ANDREW).

A MONK OF FIFE ; a Story of the Days of Joan of Arc. With 13 Illustrations by SELWYN IMAGE. Crown 8vo, 3s. 6d.

THE DISENTANGLERS. With 7 Full-page Illustrations by H. J. FORD. Crown 8vo, 6s.

Lyall (EDNA).

THE HINDERERS. Crown 8vo, 2s. 6d.

THE AUTOBIOGRAPHY OF A SLANDER. Fcp. 8vo, 1s. sewed.

Presentation Edition. With 20 Illustrations by LANCELOT SPEED. Cr. 8vo, 2s. 6d. net.

DOREEN. The Story of a Singer. Crown 8vo, 6s.

WAYFARING MEN. Crown 8vo, 6s.

HOPE THE HERMIT : a Romance of Borrowdale. Crown 8vo, 6s.

Marchmont.—IN THE NAME OF A WOMAN : a Romance. By ARTHUR W. MARCHMONT. With 8 Illustrations. Crown 8vo, 6s.

Mason and Lang.—P A R S O N KELLY. By A. E. W. MASON and ANDREW LANG. Crown 8vo, 3s. 6d.

Max Müller.—DEUTSCHE LIEBE (GERMAN LOVE) : Fragments from the Papers of an Alien. Collected by F. MAX MÜLLER. Translated from the German by G. A. M. Crown 8vo, gilt top, 5s.

Melville (G. J. WHYTE).

The Gladiators.   Holmby House.
The Interpreter.   Kate Coventry.
Good for Nothing.   Digby Grand.
The Queen's Maries.   General Bounce.

Crown 8vo, 1s. 6d. each.

Merriman.—FLOTSAM : A Story of the Indian Mutiny. By HENRY SETON MERRIMAN. With Frontispiece and Vignette by H. G. MASSEY. Crown 8vo, 3s. 6d.

Morris (WILLIAM).

THE SUNDERING FLOOD. Crown 8vo, 7s. 6d.

THE WATER OF THE WONDROUS ISLES. Crown 8vo, 7s. 6d.

THE WELL AT THE WORLD'S END. 2 vols. 8vo, 28s.

THE WOOD BEYOND THE WORLD. Crown 8vo, 6s. net.

THE STORY OF THE GLITTERING PLAIN, which has been also called The Land of the Living Men, or The Acre of the Undying. Square post 8vo, 5s. net.

THE ROOTS OF THE MOUNTAINS, wherein is told somewhat of the Lives of the Men of Burgdale, their Friends, their Neighbours, their Foemen, and their Fellows-in-Arms. Written in Prose and Verse. Square cr. 8vo, 8s.

A TALE OF THE HOUSE OF THE WOLFINGS, and all the Kindreds of the Mark. Written in Prose and Verse. Square crown 8vo, 6s.

A DREAM OF JOHN BALL. AND A KING'S LESSON. 12mo, 1s. 6d.

NEWS FROM NOWHERE : or, An Epoch of Rest. Being some Chapters from an Utopian Romance. Post 8vo, 1s. 6d.

THE STORY OF GRETTIR THE STRONG. Translated from the Icelandic by EIRÍKR MAGNÚSSON and WILLIAM MORRIS. Crown 8vo, 5s. net.

THREE NORTHERN LOVE STORIES, and other Tales. Translated from the Icelandic by EIRÍKR MAGNÚSSON and WILLIAM MORRIS. Crown 8vo, 6s. net.

** For Mr. William Morris's other Works, see pp. 24, 37 and 40.

Fiction, Humour, etc.—*continued.*

Newman (CARDINAL).

LOSS AND GAIN : The Story of a Convert. Crown 8vo, 3s. 6d.

CALLISTA : a Tale of the Third Century. Crown 8vo, 3s. 6d.

Phillipps-Wolley.—SNAP : A Legend of the Lone Mountain. By C. PHILLIPPS-WOLLEY. With 13 Illustrations. Crown 8vo, 3s. 6d.

Ridley.—ANNE MAINWARING. By ALICE RIDLEY, Author of 'The Story of Aline'. Crown 8vo, 6s.

Sewell (ELIZABETH M.).

A Glimpse of the World.	Amy Herbert.
Laneton Parsonage.	Cleve Hall.
Margaret Percival.	Gertrude.
Katherine Ashton.	Home Life.
The Earl's Daughter.	After Life.
The Experience of Life.	Ursula. Ivors.

Crown 8vo, cloth plain, 1s. 6d. each ; cloth extra, gilt edges, 2s. 6d. each.

Sheehan.—LUKE DELMEGE. By the Rev. P. A. SHEEHAN, P.P., Author of ' My New Curate '. Crown 8vo, 6s.

Somerville (E. Œ.) **and Ross** (MARTIN).

SOME EXPERIENCES OF AN IRISH R.M. With 31 Illustrations by E. Œ. SOMERVILLE. Crown 8vo, 6s.

THE REAL CHARLOTTE. Crown 8vo, 3s. 6d.

THE SILVER FOX. Crown 8vo, 3s. 6d.

Stebbing. — RACHEL WULFSTAN, and other Stories. By W. STEBBING, author of ' Probable Tales '. Crown 8vo, 4s. 6d.

Stevenson (ROBERT LOUIS).

THE STRANGE CASE OF DR. JEKYLL AND MR. HYDE. Fcp. 8vo, 1s. sewed, 1s. 6d. cloth.

THE STRANGE CASE OF DR. JEKYLL AND MR. HYDE, WITH OTHER FABLES. Cr. 8vo, bound in buckram, with gilt top, 5s. net.
' Silver Library' Edition. Crown 8vo, 3s. 6d.

MORE NEW ARABIAN NIGHTS —THE DYNAMITER. By ROBERT LOUIS STEVENSON and FANNY VAN DE GRIFT STEVENSON. Crown 8vo, 3s. 6d.

THE WRONG BOX. By ROBERT LOUIS STEVENSON and LLOYD OSBOURNE. Crown 8vo, 3s. 6d.

Suttner.—LAY DOWN YOUR ARMS (*Die Waffen Nieder*) : The Autobiography of Martha von Tilling. By BERTHA VON SUTTNER. Translated by T. HOLMES. Crown 8vo, 1s. 6d.

Swan.—BALLAST. By MYRA SWAN. Crown 8vo, 6s.

Trollope (ANTHONY).

THE WARDEN. Crown 8vo, 1s. 6d.

BARCHESTER TOWERS. Crown 8vo, 1s. 6d.

Walford (L. B.).

CHARLOTTE. Crown 8vo, 6s.

ONE OF OURSELVES. Cr. 8vo, 6s.

THE INTRUDERS. Cr. 8vo, 2s. 6d.

LEDDY MARGET. Cr. 8vo, 2s. 6d.

IVA KILDARE : a Matrimonial Problem. Crown 8vo, 2s. 6d.

MR. SMITH : a Part of his Life. Cr. 8vo, 2s. 6d.

THE BABY'S GRANDMOTHER. Crown 8vo, 2s. 6d.

Fiction, Humour, etc.—*continued.*

Walford (L. B.)—*continued.*

COUSINS. Crown 8vo, 2s. 6d.

TROUBLESOME DAUGHTERS. Cr. 8vo, 2s. 6d.

PAULINE. Crown 8vo, 2s. 6d.

DICK NETHERBY. Cr. 8vo, 2s. 6d.

THE HISTORY OF A WEEK. Cr. 8vo, 2s. 6d.

A STIFF-NECKED GENERATION. Crown 8vo, 2s. 6d.

NAN, and other Stories. Crown 8vo, 2s. 6d.

THE MISCHIEF OF MONICA. Cr. 8vo, 2s. 6d.

THE ONE GOOD GUEST. Crown 8vo, 2s. 6d.

'PLOUGHED,' and other Stories. Cr. 8vo, 2s. 6d.

THE MATCHMAKER. Crown 8vo, 2s. 6d.

Ward.—ONE POOR SCRUPLE. By Mrs. WILFRID WARD. Crown 8vo, 6s.

Weyman (STANLEY).

THE HOUSE OF THE WOLF. With Frontispiece and Vignette. Crown 8vo, 3s. 6d.

A GENTLEMAN OF FRANCE. With Frontispiece and Vignette. Crown 8vo, 6s.

THE RED COCKADE. With Frontispiece and Vignette. Crown 8vo, 6s.

SHREWSBURY. With 24 Illustrations by CLAUDE A. SHEPPERSON. Cr. 8vo, 6s.

SOPHIA. With Frontispiece. Crown 8vo, 6s.

Yeats (S. LEVETT).

THE CHEVALIER D'AURIAC. Cr. 8vo, 3s. 6d.

THE TRAITOR'S WAY. Crown 8vo, 6s.

Yoxall.—THE ROMMANY STONE. By J. H. YOXALL, M.P. Crown 8vo, 6s.

Popular Science (Natural History, etc.).

Butler.—OUR HOUSEHOLD IN-SECTS. An Account of the Insect-Pests found in Dwelling-Houses. By EDWARD A. BUTLER, B.A., B.Sc. (Lond.). With 113 Illustrations. Cr. 8vo, 3s. 6d.

Furneaux (W.).

THE OUTDOOR WORLD; or, The Young Collector's Handbook. With 18 Plates (16 of which are coloured), and 549 Illustrations in the Text. Crown 8vo, gilt edges, 6s. net.

Furneaux (W.) *continued.*

BUTTERFLIES AND MOTHS (British). With 12 coloured Plates and 241 Illustrations in the Text. Crown 8vo, gilt edges, 6s. net.

LIFE IN PONDS AND STREAMS. With 8 coloured Plates and 331 Illustrations in the Text. Cr. 8vo, gilt edges, 6s. net.

Popular Science (Natural History, etc.)—*continued.*

Hartwig (GEORGE).

THE SEA AND ITS LIVING WON-DERS. With 12 Plates and 303 Woodcuts. 8vo, gilt top, 7s. net.

THE TROPICAL WORLD. With 8 Plates and 172 Woodcuts. 8vo, gilt top, 7s. net.

THE POLAR WORLD. With 3 Maps, 8 Plates and 85 Woodcuts. 8vo, gilt top, 7s. net.

THE SUBTERRANEAN WORLD. With 3 Maps and 80 Woodcuts. 8vo, gilt top, 7s. net.

Helmholtz.—POPULAR LECTURES ON SCIENTIFIC SUBJECTS. By HERMANN VON HELMHOLTZ. With 68 Woodcuts. 2 vols. Cr. 8vo, 3s. 6d. each.

Hudson (W. H.).

BIRDS AND MAN. Large Crown 8vo, 6s. net.

NATURE IN DOWNLAND. With 12 Plates and 14 Illustrations in the Text, by A. D. McCORMICK. 8vo, 10s. 6d. net.

BRITISH BIRDS. With a Chapter on Structure and Classification by FRANK E. BEDDARD, F.R.S. With 16 Plates (8 of which are Coloured), and over 100 Illustrations in the Text. Crown 8vo, gilt edges, 6s. net.

Millais.—THE NATURAL HISTORY OF THE BRITISH SURFACE-FEED-ING DUCKS. By JOHN GUILLE MILLAIS, F.Z.S., etc. With 6 Photo-gravures and 66 Plates (41 in Colours) from Drawings by the Author, ARCHI-BALD THORBURN, and from Photographs. Royal 4to, £6 6s.

Proctor (RICHARD A.).

LIGHT SCIENCE FOR LEISURE HOURS. Familiar Essays on Scien-tific Subjects. Crown 8vo, 3s. 6d.

ROUGH WAYS MADE SMOOTH. Familiar Essays on Scientific Subjects. Crown 8vo, 3s. 6d.

Proctor (RICHARD A.)—*continued.*

PLEASANT WAYS IN SCIENCE. Crown 8vo, 3s. 6d.

NATURE STUDIES. By R. A. PROC-TOR, GRANT ALLEN, A. WILSON, T. FOSTER and E. CLODD. Cr. 8vo, 3s. 6d.

LEISURE READINGS. By R. A. PROCTOR, E. CLODD, A. WILSON, T. FOSTER and A. C. RANYARD. Crown 8vo, 3s. 6d.

** *For Mr. Proctor's other books see pp. 16 and 35 and Messrs. Longmans & Co.'s Catalogue of Scientific Works.*

Stanley.—A FAMILIAR HISTORY OF BIRDS. By E. STANLEY, D.D., formerly Bishop of Norwich. With 160 Illustrations. Crown 8vo, 3s. 6d.

Wood (Rev. J. G.).

HOMES WITHOUT HANDS: A De-scription of the Habitations of Animals, classed according to their Principle of Construction. With 140 Illustrations. 8vo, gilt top, 7s. net.

INSECTS AT HOME: A Popular Account of British Insects, their Structure, Habits and Transforma-tions. With 700 Illustrations. 8vo, gilt top, 7s. net.

INSECTS ABROAD: A Popular Ac-count of Foreign Insects, their Structure, Habits and Transforma-tions. With 600 Illustrations. 8vo, 7s. net.

OUT OF DOORS: a Selection of Original Articles on Practical Natural History. With 11 Illustrations. Cr. 8vo, 3s. 6d.

PETLAND REVISITED. With 33 Illustrations. Crown 8vo, 3s. 6d.

STRANGE DWELLINGS: a Descrip-tion of the Habitations of Animals, abridged from ' Homes without Hands '. With 60 Illustrations. Cr. 8vo, 3s. 6d.

Works of Reference.

Gwilt.—AN ENCYCLOPÆDIA OF ARCHITECTURE. By JOSEPH GWILT, F.S.A. With 1700 Engravings. Revised (1888), with alterations and Considerable Additions by WYATT PAPWORTH. 8vo, 21s. net.

Longmans' GAZETTEER OF THE WORLD. Edited by GEORGE G. CHISHOLM, M.A., B.Sc. Imperial 8vo, 18s. net; cloth, 21s. half-morocco.

Maunder (SAMUEL).

BIOGRAPHICAL TREASURY. With Supplement brought down to 1889. By Rev. JAMES WOOD. Fcp. 8vo, 6s.

THE TREASURY OF BIBLE KNOWLEDGE. By the Rev. J. AYRE, M.A. With 5 Maps, 15 Plates, and 300 Woodcuts. Fcp. 8vo, 6s.

TREASURY OF KNOWLEDGE AND LIBRARY OF REFERENCE. Fcp. 8vo, 6s.

Maunder (SAMUEL)—*continued.*

THE TREASURY OF BOTANY. Edited by J. LINDLEY, F.R.S., and T. MOORE, F.L.S. With 274 Woodcuts and 20 Steel Plates. 2 vols. Fcp. 8vo, 12s.

Roget.—THESAURUS OF ENGLISH WORDS AND PHRASES. Classified and Arranged so as to Facilitate the Expression of Ideas and assist in Literary Composition. By PETER MARK ROGET, M.D., F.R.S. Recomposed throughout, enlarged and improved, partly from the Author's Notes, and with a full Index, by the Author's Son, JOHN LEWIS ROGET. Crown 8vo, 9s. net.

Willich.—POPULAR TABLES for giving information for ascertaining the value of Lifehold, Leasehold, and Church Property, the Public Funds, etc. By CHARLES M. WILLICH. Edited by H. BENCE JONES. Crown 8vo, 10s. 6d.

Children's Books.

Adelborg. — CLEAN PETER AND THE CHILDREN OF GRUBBYLEA. By OTTILIA ADELBORG. Translated from the Swedish by Mrs. GRAHAM WALLAS. With 23 Coloured Plates. Oblong 4to, boards, 3s. 6d. net.

Alick's Adventures. By G. R. With 8 Illustrations by JOHN HASSALL. Crown 8vo, 3s. 6d.

Brown.—THE BOOK OF SAINTS AND FRIENDLY BEASTS. By ABBIE FARWELL BROWN. With 8 Illustrations by FANNY Y. CORY. Cr. 8vo, 4s. 6d. net.

Buckland.—TWO LITTLE RUNAWAYS. Adapted from the French of LOUIS DESNOYERS. By JAMES BUCKLAND. With 110 Illustrations by CECIL ALDIN. Crown 8vo, 6s.

Crake (Rev. A. D.).

EDWY THE FAIR; or, The First Chronicle of Æscendune. Crown 8vo, silver top, 2s. net.

ALFGAR THE DANE: or, The Second Chronicle of Æscendune. Crown 8vo, silver top, 2s. net.

THE RIVAL HEIRS: being the Third and last Chronicle of Æscendune. Crown 8vo, silver top, 2s. net.

THE HOUSE OF WALDERNE. A Tale of the Cloister and the Forest in the Days of the Barons' Wars. Cr. 8vo, silver top, 2s. net.

BRIAN FITZ-COUNT. A Story of Wallingford Castle and Dorchester Abbey. Crown 8vo, silver top, 2s. net.

Children's Books—*continued.*

Henty (G. A.).—Edited by.

YULE LOGS: A Story Book for Boys. By VARIOUS AUTHORS. With 61 Illustrations. Cr. 8vo, gilt edges, 3s. net.

YULE-TIDE YARNS: a Story Book for Boys. By VARIOUS AUTHORS. With 45 Illustrations. Crown 8vo, gilt edges, 3s. net.

Lang (ANDREW).—Edited by.

THE VIOLET FAIRY BOOK. With 8 Coloured Plates and 54 other Illustrations. Crown 8vo, gilt edges, 6s.

THE BLUE FAIRY BOOK. With 138 Illustrations. Cr. 8vo, gilt edges, 6s.

THE RED FAIRY BOOK. With 100 Illustrations. Cr. 8vo, gilt edges, 6s.

THE GREEN FAIRY BOOK. With 99 Illustrations. Crown 8vo, gilt edges, 6s.

THE GREY FAIRY BOOK. With 65 Illustrations. Cr. 8vo, gilt edges, 6s.

THE YELLOW FAIRY BOOK. With 104 Illustrations. Crown 8vo, gilt edges, 6s.

THE PINK FAIRY BOOK. With 67 Illustrations. Crown 8vo, gilt edges, 6s.

THE BLUE POETRY BOOK. With 100 Illustrations. Crown 8vo, gilt edges, 6s.

THE TRUE STORY BOOK. With 66 Illustrations. Crown 8vo, gilt edges, 6s.

THE RED TRUE STORY BOOK. With 100 Illustrations. Cr. 8vo, gilt edges, 6s.

THE ANIMAL STORY BOOK. With 67 Illustrations. Crown 8vo, gilt edges, 6s.

THE RED BOOK OF ANIMAL STORIES. With 65 Illustrations. Crown 8vo, gilt edges, 6s.

THE ARABIAN NIGHTS ENTERTAINMENTS. With 66 Illustrations. Crown 8vo, gilt edges, 6s.

THE BOOK OF ROMANCE. With 8 Coloured Plates and 44 other Illustrations. Cr. 8vo, gilt edges, 6s.

Lyall.—THE BURGES LETTERS: a Record of Child Life in the Sixties. By EDNA LYALL. With Coloured Frontispiece and 8 other full-page Illustrations by WALTER S. STACEY. Crown 8vo, 2s. 6d.

Meade (L. T.).

DADDY'S BOY. With 8 Illustrations. Crown 8vo, gilt edges, 3s. net.

DEB AND THE DUCHESS. With 7 Illustrations. Crown 8vo, gilt edges, 3s. net.

THE BERESFORD PRIZE. With 7 Illustrations. Crown 8vo, gilt edges, 3s. net.

THE HOUSE OF SURPRISES. With 6 Illustrations. Crown 8vo, gilt edges, 3s. net.

Murray.—FLOWER LEGENDS FOR CHILDREN. By HILDA MURRAY (the Hon. Mrs. MURRAY of Elibank). Pictured by J. S. ELAND. With numerous Coloured and other Illustrations. Oblong 4to, 6s.

Penrose.—CHUBBY: A NUISANCE. By Mrs. PENROSE. With 8 Illustrations by G. G. MANTON. Crown 8vo, 3s. 6d.

Praeger (ROSAMOND).

THE ADVENTURES OF THE THREE BOLD BABES: HECTOR, HONORIA AND ALISANDER. A Story in Pictures. With 24 Coloured Plates and 24 Outline Pictures. Oblong 4to, 3s. 6d.

THE FURTHER DOINGS OF THE THREE BOLD BABES. With 24 Coloured Pictures and 24 Outline Pictures. Oblong 4to, 3s. 6d.

Roberts.—THE ADVENTURES OF CAPTAIN JOHN SMITH: Captain of Two Hundred and Fifty Horse, and sometime President of Virginia. By E. P. ROBERTS. With 17 Illustrations and 3 Maps. Crown 8vo, 5s. net.

Stevenson.—A CHILD'S GARDEN OF VERSES. By ROBERT LOUIS STEVENSON. Fcp. 8vo, gilt top, 5s.

Tappan.—OLD BALLADS IN PROSE. By EVA MARCH TAPPAN. With 4 Illustrations by FANNY Y. CORY. Crown 8vo, gilt top, 4s. 6d. net.

Children's Books—*continued.*

Upton (FLORENCE K. and BERTHA).

THE ADVENTURES OF TWO DUTCH DOLLS AND A 'GOLLIWOGG'. With 31 Coloured Plates and numerous Illustrations in the Text. Oblong 4to, 6s.

THE GOLLIWOGG'S BICYCLE CLUB. With 31 Coloured Plates and numerous Illustrations in the Text. Oblong 4to, 6s.

THE GOLLIWOGG AT THE SEA-SIDE. With 31 Coloured Plates and numerous Illustrations in the Text. Oblong 4to, 6s.

THE GOLLIWOGG IN WAR. With 31 Coloured Plates. Oblong 4to, 6s.

THE GOLLIWOGG'S POLAR AD-VENTURES. With 31 Coloured Plates. Oblong 4to, 6s.

Upton (FLORENCE K. and BERTHA)—*continued.*

THE GOLLIWOGG'S AUTO-GO-CART. With 31 Coloured Plates and numerous Illustrations in the Text. Oblong 4to, 6s.

THE GOLLIWOGG'S AIR-SHIP. With 31 Coloured Pictures and numerous Illustrations in the Text. Oblong 4to, 6s.

THE VEGE-MEN'S REVENGE. With 31 Coloured Plates and numerous Illustrations in the Text. Oblong 4to, 6s.

Wemyss.—'THINGS WE THOUGHT OF': Told from a Child's Point of View. By MARY C. E. WEMYSS, Author of ' All About All of Us '. With 8 Illustrations in Colour by S. R. PRAEGER. Crown 8vo, 3s. 6d.

THE SILVER LIBRARY.

Crown 8vo, 3s. 6d. EACH VOLUME.

Arnold's (Sir Edwin) Seas and Lands. With 17 Illustrations. 3s. 6d.

Bagehot's (W.) Biographical Studies. 3s. 6d.

Bagehot's (W.) Economic Studies. 3s. 6d.

Bagehot's (W.) Literary Studies. With Portrait. 3 vols. 3s. 6d. each.

Baker's (Sir S. W.) Eight Years in Ceylon. With 6 Illustrations. 3s. 6d.

Baker's (Sir S. W.) Rifle and Hound in Ceylon. With 6 Illustrations. 3s. 6d.

Baring-Gould's (Rev. S.) Curious Myths of the Middle Ages. 3s. 6d.

Baring-Gould's (Rev. S.) Origin and Development of Religious Belief. 2 vols. 3s. 6d. each.

Becker's W. A.) Gallus: or, Roman Scenes in the time of Augustus. With 26 Illustra-tions. 3s. 6d.

Becker's (W. A.) Charicles: or, Illustra-tions of the Private Life of the Ancient Greeks. With 26 Illustrations. 3s. 6d.

Gent's (J. T.) The Ruined Cities of Ma-shonaland. With 117 Illustrations. 3s. 6d.

Brassey's (Lady) A Voyage in the 'Sun-beam '. With 66 Illustrations. 3s. 6d.

Buckle's (H. T.) History of Civilisation in England. 3 vols. 10s. 6d.

Churchill's (Winston S.) The Story of the Malakand Field Force, 1897. With 6 Maps and Plans. 3s. 6d.

Clodd's (E.) Story of Creation: a Plain Account of Evolution. With 77 Illus-trations. 3s. 6d.

Conybeare (Rev. W. J.) and Howson's (Very Rev. J. S.) Life and Epistles of St. Paul. With 46 Illustrations. 3s. 6d.

Dougall's (L.) Beggars All; a Novel. 3s. 6d.

Doyle's (Sir A. Conan) Micah Clarke. A Tale of Monmouth's Rebellion. With 10 Illustrations. 3s. 6d.

Doyle's (Sir A. Conan) The Captain of the Polestar, and other Tales. 3s. 6d.

Doyle's (Sir A. Conan) The Refugees : A Tale of the Huguenots. With 25 Il-lustrations. 3s. 6d.

Doyle's (Sir A. Conan) The Stark Munro Letters. 3s. 6d.

Froude's (J. A.) The History of England, from the Fall of Wolsey to the Defeat of the Spanish Armada. 12 vols. 3s. 6d. each.

THE SILVER LIBRARY—*continued.*

Froude's (J. A.) The English in Ireland. 3 vols. 10s. 6d.

Froude's (J. A.) The Divorce of Catherine of Aragon. 3s. 6d.

Froude's (J. A.) The Spanish Story of the Armada, and other Essays. 3s. 6d.

Froude's (J. A.) English Seamen in the Sixteenth Century. 3s. 6d.

Froude's (J. A.) Short Studies on Great Subjects. 4 vols. 3s. 6d. each.

Froude's (J. A.) Oceana, or England and her Colonies. With 9 Illustrations. 3s. 6d.

Froude's (J. A.) The Council of Trent. 3s. 6d.

Froude's (J. A.) The Life and Letters of Erasmus. 3s. 6d.

Froude's (J. A.) Thomas Carlyle: a History of his Life.
1795-1835. 2 vols. 7s.
1834-1881. 2 vols. 7s.

Froude's (J. A.) Cæsar: a Sketch. 3s. 6d.

Froude's (J. A.) The Two Chiefs of Dunboy: an Irish Romance of the Last Century. 3s. 6d.

Froude's (J. A.) Writings, Selections from. 3s. 6d.

Gleig's (Rev. G. R.) Life of the Duke of Wellington. With Portrait. 3s. 6d.

Greville's (C. C. F.) Journal of the Reigns of King George IV., King William IV., and Queen Victoria. 8 vols. 3s. 6d. each.

Haggard's (H. R.) She: A History of Adventure. With 32 Illustrations. 3s. 6d.

Haggard's (H. R.) Allan Quatermain. With 20 Illustrations. 3s. 6d.

Haggard's (H. R.) Colonel Quaritch, V.C.: a Tale of Country Life. With Frontispiece and Vignette. 3s. 6d.

Haggard's (H. R.) Cleopatra. With 29 Illustrations. 3s. 6d.

Haggard's (H. R.) Eric Brighteyes. With 51 Illustrations. 3s. 6d.

Haggard's (H. R.) Beatrice. With Frontispiece and Vignette. 3s. 6d.

Haggard's (H. R.) Allan's Wife. With 34 Illustrations. 3s. 6d.

Haggard's (H. R.) Heart of the World. With 15 Illustrations. 3s. 6d.

Haggard's (H. R.) Montezuma's Daughter. With 25 Illustrations. 3s. 6d.

Haggard's (H. R.) Swallow: a Tale of the Great Trek. With 8 Illustrations. 3s. 6d.

Haggard's (H. R.) The Witch's Head. With 16 Illustrations. 3s. 6d.

Haggard's (H. R.) Mr. Meeson's Will. With 16 Illustrations. 3s. 6d.

Haggard's (H. R.) Nada the Lily. With 23 Illustrations. 3s. 6d.

Haggard's (H. R.) Dawn. With 16 Illustrations. 3s. 6d.

Haggard's (H. R.) The People of the Mist. With 16 Illustrations. 3s. 6d.

Haggard's (H. R.) Joan Haste. With 20 Illustrations. 3s. 6d.

Haggard (H. R.) and Lang's (A.) The World's Desire. With 27 Illus. 3s. 6d.

Harte's (Bret) In the Carquinez Woods, and other Stories. 3s. 6d.

Helmholtz's (Hermann von) Popular Lectures on Scientific Subjects. With 68 Illustrations. 2 vols. 3s. 6d. each.

Hope's (Anthony) The Heart of Princess Osra. With 9 Illustrations. 3s. 6d.

Howitt's (W.) Visits to Remarkable Places. With 80 Illustrations. 3s. 6d.

Jefferies' (R.) The Story of My Heart: My Autobiography. With Portrait. 3s. 6d.

Jefferies' (R.) Field and Hedgerow. With Portrait. 3s. 6d.

Jefferies' (R.) Red Deer. With 17 Illustrations. 3s. 6d.

Jefferies' (R.) Wood Magic: a Fable. With Frontispiece and Vignette by E. V. B. 3s. 6d.

Jefferies' (R.) The Toilers of the Field. With Portrait from the Bust in Salisbury Cathedral. 3s. 6d.

Kaye (Sir J.) and Malleson's (Colonel) History of the Indian Mutiny of 1857-8. 6 vols. 3s. 6d. each.

Knight's (E. F.) The Cruise of the 'Alerte': the Narrative of a Search for Treasure on the Desert Island of Trinidad. With 2 Maps and 23 Illustrations. 3s. 6d.

Knight's (E. F.) Where Three Empires Meet: a Narrative of Recent Travel in Kashmir, Western Tibet, Baltistan, Gilgit. With a Map and 54 Illustrations. 3s. 6d.

THE SILVER LIBRARY—*continued.*

Knight's (E. F.) The 'Falcon' on the Baltic : a Coasting Voyage from Hammersmith to Copenhagen in a Three-Ton Yacht. With Map and 11 Illustrations. 3s. 6d.

Köstlin's (J.) Life of Luther. With 62 Illustrations and 4 Facsimiles of MSS. 3s. 6d.

Lang's (A.) Angling Sketches. With 20 Illustrations. 3s. 6d.

Lang's (A.) Custom and Myth: Studies of Early Usage and Belief. 3s. 6d.

Lang's (A.) Cock Lane and Common-Sense. 3s. 6d.

Lang's (A.) The Book of Dreams and Ghosts. 3s. 6d.

Lang's (A.) A Monk of Fife: a Story of the Days of Joan of Arc. With 13 Illustrations. 3s. 6d.

Lang's (A.) Myth, Ritual and Religion. 2 vols. 7s.

Lees (J. A.) and Clutterbuck's (W.J.) B.C. 1887. A Ramble in British Columbia. With Maps and 75 Illustrations. 3s. 6d.

Levett-Yeats' (S.) The Chevalier D'Auriac. 3s. 6d.

Macaulay's (Lord) Complete Works. 'Albany' Edition. With 12 Portraits. 12 vols. 3s. 6d. each.

Macaulay's (Lord) Essays and Lays of Ancient Rome, etc. With Portrait and 4 Illustrations to the 'Lays'. 3s. 6d.

Macleod's (H. D.) Elements of Banking. 3s. 6d.

Marshman's (J. C.) Memoirs of Sir Henry Havelock. 3s. 6d.

Mason (A. E. W.) and Lang's (A.) Parson Kelly. 3s. 6d.

Merivale's (Dean) History of the Romans under the Empire. 8 vols. 3s. 6d. each.

Merriman's (H. S.) Flotsam : a Tale of the Indian Mutiny. 3s. 6d.

Mill's (J. S.) Political Economy. 3s. 6d.

Mill's (J. S.) System of Logic. 3s. 6d.

Milner's (Geo.) Country Pleasures: the Chronicle of a year chiefly in a Garden. 3s. 6d.

Nansen's (F.) The First Crossing of Greenland. With 142 Illustrations and a Map. 3s. 6d.

Phillipps-Wolley's (C.) Snap: a Legend of the Lone Mountain. With 13 Illustrations. 3s. 6d.

Proctor's (R. A.) The Orbs Around Us. 3s. 6d.

Proctor's (R. A.) The Expanse of Heaven. 3s. 6d.

Proctor's (R. A.) Light Science for Leisure Hours. 3s. 6d.

Proctor's (R. A.) The Moon. 3s. 6d.

Proctor's (R. A.) Other Worlds than Ours. 3s. 6d.

Proctor's (R. A.) Our Place among Infinities: a Series of Essays contrasting our Little Abode in Space and Time with the Infinities around us. 3s. 6d.

Proctor's (R. A.) Other Suns than Ours. 3s. 6d.

Proctor's (R. A.) Rough Ways made Smooth. 3s. 6d.

Proctor's (R. A.) Pleasant Ways in Science. 3s. 6d.

Proctor's (R. A.) Myths and Marvels of Astronomy. 3s. 6d.

Proctor's (R. A.) Nature Studies. 3s. 6d.

Proctor's (R. A.) Leisure Readings. By R. A. PROCTOR, EDWARD CLODD, ANDREW WILSON, THOMAS FOSTER and A. C. RANYARD. With Illustrations. 3s. 6d.

Rossetti's (Maria F.) A Shadow of Dante. 3s. 6d.

Smith's (R. Bosworth) Carthage and the Carthaginians. With Maps, Plans, etc. 3s. 6d.

Stanley's (Bishop) Familiar History of Birds. With 160 Illustrations. 3s. 6d.

Stephen's (Sir L.) The Playground of Europe (The Alps). With 4 Illustrations. 3s. 6d.

Stevenson's (R. L.) The Strange Case of Dr. Jekyll and Mr. Hyde ; with other Fables. 3s. 6d.

Stevenson (R. L.) and Osbourne's (Ll.) The Wrong Box. 3s. 6d.

Stevenson (Robt. Louis) and Stevenson's (Fanny van de Grift) More New Arabian Nights. The Dynamiter. 3s. 6d.

Trevelyan's (Sir G. O.) The Early History of Charles James Fox. 3s. 6d.

Weyman's (Stanley J.) The House of the Wolf: a Romance. 3s. 6d.

Wood's (Rev. J. G.) Petland Revisited. With 33 Illustrations. 3s. 6d.

Wood's (Rev. J. G.) Strange Dwellings. With 60 Illustrations. 3s. 6d.

Wood's (Rev. J. G.) Out of Doors. With 11 Illustrations. 3s. 6d.

Cookery, Domestic Management, etc.

Acton.—MODERN COOKERY. By ELIZA ACTON. With 150 Woodcuts. Fcp. 8vo, 4s. 6d.

Angwin. -- SIMPLE HINTS ON CHOICE OF FOOD, with Tested and Economical Recipes. For Schools, Homes and Classes for Technical Instruction. By M. C. ANGWIN, Diplomate (First Class) of the National Union for the Technical Training of Women, etc. Crown 8vo, 1s.

Ashby.—HEALTH IN THE NURSERY. By HENRY ASHBY, M.D., F.R.C.P., Physician to the Manchester Children's Hospital. With 25 Illustrations. Cr. 8vo, 3s. net.

Bull (THOMAS, M.D.).
HINTS TO MOTHERS ON THE MANAGEMENT OF THEIR HEALTH DURING THE PERIOD OF PREGNANCY. Fcp. 8vo, sewed, 1s. 6d. ; cloth, gilt edges, 2s. net.
THE MATERNAL MANAGEMENT OF CHILDREN IN HEALTH AND DISEASE. Fcp. 8vo, sewed, 1s. 6d. ; cloth, gilt edges, 2s. net.

De Salis (MRS.).
A LA MODE COOKERY : UP-TO-DATE RECIPES. With 24 Plates (16 in Colours). Crown 8vo, 5s. net.
CAKES AND CONFECTIONS À LA MODE. Fcp. 8vo, 1s. 6d.
DOGS : A Manual for Amateurs. Fcp. 8vo, 1s. 6d.
DRESSED GAME AND POULTRY À LA MODE. Fcp. 8vo, 1s. 6d.
DRESSED VEGETABLES À LA MODE. Fcp. 8vo, 1s. 6d.
DRINKS À LA MODE. Fcp. 8vo, 1s. 6d.

De Salis (MRS.)—*continued.*
ENTRÉES À LA MODE. Fcp. 8vo, 1s. 6d.
FLORAL DECORATIONS. Fcp. 8vo, 1s. 6d.
GARDENING À LA MODE. Fcp. 8vo, Part I., Vegetables, 1s. 6d. Part II., Fruits, 1s. 6d.
NATIONAL VIANDS À LA MODE. Fcp. 8vo, 1s. 6d.
NEW-LAID EGGS. Fcp. 8vo, 1s. 6d.
OYSTERS À LA MODE. Fcp. 8vo, 1s. 6d.
PUDDINGS AND PASTRY À LA MODE. Fcp. 8vo, 1s. 6d.
SAVOURIES À LA MODE. Fcp. 8vo, 1s. 6d.
SOUPS AND DRESSED FISH À LA MODE. Fcp. 8vo, 1s. 6d.
SWEETS AND SUPPER DISHES À LA MODE. Fcp. 8vo, 1s. 6d.
TEMPTING DISHES FOR SMALL INCOMES. Fcp. 8vo, 1s. 6d.
WRINKLES AND NOTIONS FOR EVERY HOUSEHOLD. Crown 8vo, 1s. 6d.

Lear.—MAIGRE COOKERY. By H. L. SIDNEY LEAR. 16mo, 2s.

Poole.—COOKERY FOR THE DIABETIC. By W. H. and Mrs. POOLE. With Preface by Dr. PAVY. Fcp. 8vo, 2s. 6d.

Rotheram. — HOUSEHOLD COOKERY RECIPES. By M. A. ROTHERAM, First Class Diplomée, National Training School of Cookery, London ; Instructress to the Bedfordshire County Council. Crown 8vo, 2s.

The Fine Arts and Music.

Burne-Jones. - THE BEGINNING OF THE WORLD : Twenty - five Pictures by Sir EDWARD BURNE-JONES, Bart. Medium 4to, Boards, 7s. 6d. net.

Burns and Colenso. -- LIVING ANATOMY. By CECIL L. BURNS, R.B.A., and ROBERT J. COLENSO, M.A., M.D. 40 Plates, 11¼ × 8¾ in., each Plate containing Two Figures--(a) A Natural Male or Female Figure ; (b) The same Figure Anatomised. In a Portfolio. 7s. 6d. net.

Hamlin.—A TEXT-BOOK OF THE HISTORY OF ARCHITECTURE. By A. D. F. HAMLIN, A.M. With 229 Illustrations. Crown 8vo, 7s. 6d.

Haweis (Rev. H. R.).
MUSIC AND MORALS. With Portrait of the Author, and Numerous Illustrations, Facsimiles and Diagrams. Crown 8vo, 6s. net.
MY MUSICAL LIFE. With Portrait of Richard Wagner and 3 Illustrations. Crown 8vo, 6s. net.

Fine Arts and Music—*continued.*

Huish, Head and Longman.— SAMPLERS AND TAPESTRY EMBROIDERIES. By MARCUS B. HUISH, LL.B. ; also 'The Stitchery of the Same,' by Mrs. HEAD; and 'Foreign Samplers,' by Mrs. C. J. LONGMAN. With 30 Reproductions in Colour and 40 Illustrations in Monochrome. 4to, £2 2s. net.

Hullah.—THE HISTORY OF MODERN MUSIC. By JOHN HULLAH. 8vo, 8s. 6d.

Jameson (Mrs. ANNA).

SACRED AND LEGENDARY ART, containing Legends of the Angels and Archangels, the Evangelists, the Apostles, the Doctors of the Church, St. Mary Magdalene, the Patron Saints, the Martyrs, the Early Bishops, the Hermits and the Warrior-Saints of Christendom, as represented in the Fine Arts. With 19 Etchings and 187 Woodcuts. 2 vols. 8vo, 20s. net.

LEGENDS OF THE MONASTIC ORDERS, as represented in the Fine Arts, comprising the Benedictines and Augustines, and Orders derived from their rules, the Mendicant Orders, the Jesuits, and the Order of the Visitation of St. Mary. With 11 Etchings and 88 Woodcuts. 1 vol. 8vo, 10s. net.

LEGENDS OF THE MADONNA, OR BLESSED VIRGIN MARY. Devotional with and without the Infant Jesus, Historical from the Annunciation to the Assumption, as represented in Sacred and Legendary Christian Art. With 27 Etchings and 165 Woodcuts. 1 vol. 8vo, 10s. net.

THE HISTORY OF OUR LORD, as exemplified in Works of Art, with that of His Types, St. John the Baptist, and other persons of the Old and New Testament. Commenced by the late Mrs. JAMESON; continued and completed by LADY EASTLAKE. With 31 Etchings and 281 Woodcuts. 2 vols. 8vo, 20s. net.

Kristeller.—ANDREA MANTEGNA. By PAUL KRISTELLER. English Edition by S. ARTHUR STRONG, M.A., Librarian to the House of Lords, and at Chatsworth. With 26 Photogravure Plates and 162 Illustrations in the Text. 4to, gilt top, £3 10s. net.

Macfarren.—LECTURES ON HARMONY. By Sir GEORGE A. MACFARREN. 8vo, 12s.

Morris (WILLIAM).

ARCHITECTURE, INDUSTRY AND WEALTH. Collected Papers. Crown 8vo, 6s. net.

HOPES AND FEARS FOR ART. Five Lectures delivered in Birmingham, London, etc., in 1878-1881. Crown 8vo, 4s. 6d.

AN ADDRESS DELIVERED AT THE DISTRIBUTION OF PRIZES TO STUDENTS OF THE BIRMINGHAM MUNICIPAL SCHOOL OF ART ON 21ST FEBRUARY, 1894. 8vo, 2s. 6d. net. (*Printed in 'Golden' Type.*)

SOME HINTS ON PATTERN - DESIGNING : a Lecture delivered at the Working Men's College, London, on 10th December, 1881. 8vo, 2s. 6d. net. (*Printed in 'Golden' Type.*)

ARTS AND ITS PRODUCERS (1888) AND THE ARTS AND CRAFTS OF TO-DAY (1889). 8vo, 2s. 6d. net. (*Printed in 'Golden' Type.*)

ARCHITECTURE AND HISTORY, AND WESTMINSTER ABBEY. Two Papers read before the Society for the Protection of Ancient Buildings. 8vo, 2s. 6d. net. (*Printed in 'Golden' Type.*)

ARTS AND CRAFTS ESSAYS BY MEMBERS OF THE ARTS AND CRAFTS EXHIBITION SOCIETY. With a Preface by WILLIAM MORRIS. Crown 8vo, 2s. 6d. net.

. *For Mr. William Morris's other works see pp. 24, 27 and 40.*

The Fine Arts and Music—*continued.*

Robertson.—OLD ENGLISH SONGS AND DANCES. Decorated in Colour by W. GRAHAM ROBERTSON. Royal 4to, 42s. net.

Scott.—THE PORTRAITURES OF JULIUS CÆSAR: a Monograph. By FRANK J. SCOTT. With many Full-page Plates and Illustrations in the Text. Small 4to.

Vanderpoel.—COLOUR PROBLEMS: a Practical Manual for the Lay Student of Colour. By EMILY NOYES VANDER-POEL. With 117 Plates in Colour. Square 8vo, 21s. net.

Van Dyke.—A TEXT-BOOK ON THE HISTORY OF PAINTING. By JOHN C. VAN DYKE. With 110 Illustrations. Crown 8vo, 6s.

Wellington.—A DESCRIPTIVE AND HISTORICAL CATALOGUE OF THE COLLECTIONS OF PICTURES AND SCULPTURE AT APSLEY HOUSE, LONDON. By EVELYN, Duchess of Wellington. Illustrated by 52 Photo-Engravings, specially executed by BRAUN, CLÉMENT & CO., of Paris. 2 vols. Royal 4to, £6 6s. net.

Miscellaneous and Critical Works.

Annals of Mathematics (under the Auspices of Harvard University). Issued Quarterly. 4to, 2s. net each number.

Auto da Fé and other Essays: some being Essays in Fiction. By the Author of 'Essays in Paradox' and 'Exploded Ideas'. Crown 8vo, 5s.

Bagehot.—LITERARY STUDIES. By WALTER BAGEHOT. With Portrait. 3 vols. Crown 8vo, 3s. 6d. each.

Baker.—EDUCATION AND LIFE: Papers and Addresses. By JAMES H. BAKER, M.A., LL.D. Crown 8vo, 4s. 6d.

Baring-Gould.—CURIOUS MYTHS OF THE MIDDLE AGES. By Rev. S. BARING-GOULD. Crown 8vo, 3s. 6d.

Baynes.—SHAKESPEARE STUDIES, and other Essays. By the late THOMAS SPENCER BAYNES, LL.B., LL.D. With a Biographical Preface by Professor LEWIS CAMPBELL. Crown 8vo, 7s. 6d.

Bonnell. — CHARLOTTE BRONTË, GEORGE ELIOT, JANE AUSTIN: Studies in their Works. By HENRY H. BONNELL. Crown 8vo.

Booth. — THE DISCOVERY AND DECIPHERMENT OF THE TRI-LINGUAL CUNEIFORM INSCRIP-TIONS. By ARTHUR JOHN BOOTH, M.A. With a Plan of Persepolis. 8vo, 14s. net.

Charities Register, THE ANNUAL, AND DIGEST: being a Classified Re-gister of Charities in or available in the Metropolis. 8vo, 4s.

Christie.—SELECTED ESSAYS. By RICHARD COPLEY CHRISTIE, M.A. Oxon., Hon. LL.D. Vict. With 2 Por-traits and 3 other Illustrations. 8vo, 12s. net.

Dickinson.—KING ARTHUR IN CORNWALL. By W. HOWSHIP DICK-INSON, M.D. With 5 Illustrations. Crown 8vo, 4s. 6d.

Essays in Paradox. By the Author of 'Exploded Ideas' and 'Times and Days'. Crown 8vo, 5s.

Evans.—THE ANCIENT STONE IM-PLEMENTS, WEAPONS AND ORNA-MENTS OF GREAT BRITAIN. By Sir JOHN EVANS, K.C.B. With 537 Illustrations. 8vo, 28s.

Exploded Ideas, AND OTHER ESSAYS. By the Author of 'Times and Days'. Crown 8vo, 5s.

Frost.—A MEDLEY BOOK. By GEO. FROST. Crown 8vo, 3s. 6d. net.

Geikie. — THE VICAR AND HIS FRIENDS. By CUNNINGHAM GEIKIE, D.D., LL.D. Crown 8vo, 5s. net.

Haggard (H. RIDER).

A FARMER'S YEAR: being his Commonplace Book for 1898. With 36 Illustrations. Cr. 8vo, 7s. 6d. net.

RURAL ENGLAND. With 23 Agri-cultural Maps and 56 Illustrations from Photographs. 2 vols., 8vo, 3s. 6d. net.

Hoenig.—INQUIRIES CONCERNING THE TACTICS OF THE FUTURE. By FRITZ HOENIG. With 1 Sketch in the Text and 5 Maps. Translated by Captain H. M. BOWER. 8vo, 15s. net.

Miscellaneous and Critical Works—*continued.*

Hutchinson.—DREAMS AND THEIR MEANINGS. By HORACE G. HUTCHINSON. 8vo, gilt top, 9s. 6d. net.

Jefferies (RICHARD).
FIELD AND HEDGEROW. With Portrait. Crown 8vo, 3s. 6d.
THE STORY OF MY HEART: my Autobiography. Crown 8vo, 3s. 6d.
RED DEER. With 17 Illustrations. Crown 8vo, 3s. 6d.
THE TOILERS OF THE FIELD. Cr. 8vo, 3s. 6d.
WOOD MAGIC : a Fable. Crown 8vo, 3s. 6d.

Jekyll (GERTRUDE).
HOME AND GARDEN : Notes and Thoughts, Practical and Critical, of a Worker in both. With 53 Illustrations from Photographs. 8vo, 10s. 6d. net.
WOOD AND GARDEN : Notes and Thoughts, Practical and Critical, of a Working Amateur. With 71 Photographs. 8vo, 10s. 6d. net.

Johnson (J. & J. H.).
THE PATENTEE'S MANUAL : a Treatise on the Law and Practice of Letters Patent. 8vo, 10s. 6d.
AN EPITOME OF THE LAW AND PRACTICE CONNECTED WITH PATENTS FOR INVENTIONS. With a Reprint of the Patents Acts of 1883, 1885, 1886 and 1888. Crown 8vo, 2s. 6d.

Joyce.—THE ORIGIN AND HISTORY OF IRISH NAMES OF PLACES. By P. W. JOYCE, LL.D. 2 vols. Crown 8vo, 5s. each.

Lang (ANDREW).
LETTERS TO DEAD AUTHORS. Fcp. 8vo, 2s. 6d. net.
BOOKS AND BOOKMEN. With 2 Coloured Plates and 17 Illustrations. Fcp. 8vo, 2s. 6d. net.
OLD FRIENDS. Fcp. 8vo, 2s. 6d. net.
LETTERS ON LITERATURE. Fcp. 8vo, 2s. 6d. net.
ESSAYS IN LITTLE. With Portrait of the Author. Crown 8vo, 2s. 6d.
COCK LANE AND COMMON-SENSE. Crown 8vo, 3s. 6d.
THE BOOK OF DREAMS AND GHOSTS. Crown 8vo, 3s. 6d.

Maryon. — HOW THE GARDEN GREW. By MAUD MARYON. With 4 Illustrations. Crown 8vo, 5s. net.

Matthews.—NOTES ON SPEECH-MAKING. By BRANDER MATTHEWS. Fcp. 8vo, 1s. 6d. net.

Max Müller (The Right Hon. F.).
COLLECTED WORKS. 18 vols. Cr. 8vo, 5s. each.
Vol. I. NATURAL RELIGION : the Gifford Lectures, 1888.
Vol. II. PHYSICAL RELIGION : the Gifford Lectures, 1890.
Vol. III. ANTHROPOLOGICAL RELIGION : the Gifford Lectures, 1891.
Vol. IV. THEOSOPHY ; or, Psychological Religion : the Gifford Lectures, 1892.

CHIPS FROM A GERMAN WORKSHOP.
Vol. V. Recent Essays and Addresses.
Vol. VI. Biographical Essays.
Vol. VII. Essays on Language and Literature.
Vol. VIII. Essays on Mythology and Folk-lore.

Vol. IX. THE ORIGIN AND GROWTH OF RELIGION, as illustrated by the Religions of India : the Hibbert Lectures, 1878.
Vol. X. BIOGRAPHIES OF WORDS, AND THE HOME OF THE ARYAS.
Vols. XI., XII. THE SCIENCE OF LANGUAGE : Founded on Lectures delivered at the Royal Institution in 1861 and 1863. 2 vols. 10s.
Vol. XIII. INDIA : What can it Teach Us ?
Vol. XIV. INTRODUCTION TO THE SCIENCE OF RELIGION. Four Lectures, 1870.
Vol. XV. RÁMAKRISHNA : his Life and Sayings.
Vol. XVI. THREE LECTURES ON THE VEDÁNTA PHILOSOPHY, 1894.
Vol. XVII. LAST ESSAYS. First Series. Essays on Language, Folk-lore, etc.
Vol. XVIII. LAST ESSAYS. Second Series. Essays on the Science of Religion.

Miscellaneous and Critical Works—*continued.*

Milner. — COUNTRY PLEASURES: the Chronicle of a Year chiefly in a Garden. By GEORGE MILNER. Crown 8vo, 3s. 6d.

Morris.—SIGNS OF CHANGE. Seven Lectures delivered on various Occasions. By WILLIAM MORRIS. Post 8vo, 4s. 6d.

Parker and Unwin.—THE ART OF BUILDING A HOME : a Collection of Lectures and Illustrations. By BARRY PARKER and RAYMOND UNWIN. With 68 Full-page Plates. 8vo, 10s. 6d. net.

Pollock.—JANE AUSTEN : her Contemporaries and Herself. By WALTER HERRIES POLLOCK. Crown 8vo, 3s. 6d. net.

Poore (GEORGE VIVIAN, M.D.).
ESSAYS ON RURAL HYGIENE. With 13 Illustrations. Crown 8vo, 6s. 6d.

THE DWELLING HOUSE. With 36 Illustrations. Crown 8vo, 3s. 6d.

THE EARTH IN RELATION TO THE PRESERVATION AND DESTRUCTION OF CONTAGIA : being the Milroy Lectures delivered at the Royal College of Physicians in 1899, together with other Papers on Sanitation. With 13 Illustrations. Crown 8vo, 5s.

Rossetti.—A SHADOW OF DANTE : being an Essay towards studying Himself, his World, and his Pilgrimage. By MARIA FRANCESCA ROSSETTI. Crown 8vo, 3s. 6d.

Shadwell.—DRINK : TEMPERANCE AND LEGISLATION. By ARTHUR SHADWELL, M.A., M.D. Crown 8vo, 5s. net.

Soulsby (LUCY H. M.).
STRAY THOUGHTS ON READING. Fcp. 8vo, 3s. 6d. net.

STRAY THOUGHTS FOR GIRLS. 16mo, 1s. 6d. net.

STRAY THOUGHTS FOR MOTHERS AND TEACHERS. Fcp. 8vo, 2s. 6d. net.

STRAY THOUGHTS FOR INVALIDS. 16mo, 2s. net.

STRAY THOUGHTS ON CHARACTER. Fcp. 8vo, 2s. 6d. net.

25,000/11/02.

Southey.—THE CORRESPONDENCE OF ROBERT SOUTHEY WITH CAROLINE BOWLES. Edited by EDWARD DOWDEN. 8vo, 14s.

Stevens.—ON THE STOWAGE OF SHIPS AND THEIR CARGOES. With Information regarding Freights, Charter-Parties, etc. By ROBERT WHITE STEVENS. 8vo, 21s.

Sutherland. — TWENTIETH CENTURY INVENTIONS : A FORECAST. By GEORGE SUTHERLAND, M.A. Crown 8vo, 4s. 6d. net.

Thuillier.—THE PRINCIPLES OF LAND DEFENCE, AND THEIR APPLICATION TO THE CONDITIONS OF TO-DAY. By Captain H. F. THUILLIER, R.E. With Maps and Plans. 8vo, 12s. 6d. net.

Turner and Sutherland. — THE DEVELOPMENT OF AUSTRALIAN LITERATURE. By HENRY GYLES TURNER and ALEXANDER SUTHERLAND. With Portraits and Illustrations. Crown 8vo, 5s.

Warwick.—PROGRESS IN WOMEN'S EDUCATION IN THE BRITISH EMPIRE : being the Report of Conferences and a Congress held in connection with the Educational Section, Victorian Era Exhibition. Edited by the COUNTESS OF WARWICK. Crown 8vo, 6s.

Weathers.—A PRACTICAL GUIDE TO GARDEN PLANTS. By JOHN WEATHERS, F.R.H.S. With 159 Diagrams. 8vo, 21s. net.

Webb.—THE MYSTERY OF WILLIAM SHAKESPEARE: A Summary of Evidence. By his Honour Judge T. WEBB, sometime Regius Professor of Laws and Public Orator in the University of Dublin. 8vo, 10s. 6d. net.

Whittall.—FREDERIC THE GREAT ON KINGCRAFT, from the Original Manuscript; with Reminiscences and Turkish Stories. By Sir J. WILLIAM WHITTALL, President of the British Chamber of Commerce of Turkey. 8vo, 7s. 6d. net.